DETAIL DESIGN OF
MARINE SCREW PROPELLERS

PITMAN'S MECHANIC'S AND DRAUGHTSMAN'S POCKET-BOOK

BY

W. E. DOMMETT, Wh.Ex., A.M.I.A.E.

This pocket-book has been drawn up with a view to the requirements of mechanics, foremen, draughtsmen, and similar grades who have to use tables and occasionally to employ the more general mathematical problems in the course of their work.

Size 6½ in. × 4 in.
Cloth, **2s. 6d. net**

SIR ISAAC PITMAN & SONS, LTD.
1 Amen Corner, London, E.C. 4

DETAIL DESIGN OF MARINE SCREW PROPELLERS

BY

DOUGLAS H. JACKSON
M.I.Mar.E.

WITH NUMEROUS EXAMPLES AND ILLUSTRATIONS

LONDON
SIR ISAAC PITMAN & SONS, LTD., 1 Amen Corner, E.C. 4
(INCORPORATING WHITTAKER AND CO.)
BATH, MELBOURNE, AND NEW YORK
1920

PREFACE

THE full application of theory to actual design has always been rather a stumbling-block to engineers. In no branch of engineering is this more true than in the design of screw propellers. Many admirable books have dealt with the theory of screw-propeller design in a manner which appeals to the expert, but which carries theoretical considerations too far for the average designer or student. In a few cases (notable among which is Mr. Seaton's book) the more practical side of design has been covered.

In his own experience the writer has often felt the lack of a textbook which would give in a small compass an outline of the accepted theories of screw-propeller design, a full treatment of the practical application of those theories, and a short description of the various manufacturing processes. The book now presented is an attempt to satisfy the designer's need. It is in no sense a treatise, as no new theory is expounded; but it is intended to be a handy textbook for the draughtsman-designer and for the student.

In the first portion of the book the evolution of a propeller is traced out from the earliest stage of its design, through the calculation of scantlings, to the preparation of working drawings. At the end of the first portion a brief explanation is offered of the various theories and formulæ of which the application has been considered in the previous chapters. This plan, which is a reversal of the customary arrangement, has been adopted with a view to enabling the reader to concentrate upon the successive stages of design without the loss of time involved by the consideration of first principles. It is an unfortunate necessity that, in the rush of office practice, many processes of engineering design must be carried out without a full appreciation of the basic principles applied in the formulæ and theories which are used; but this custom has to be recognized in the preparation of a handbook suitable for drawing-office use.

In the second part of the book an outline treatment is given of the principal methods of pattern making and moulding, together with the machining and finishing of the propeller. To complete the scheme a chapter has been added dealing with repair work as applied to propellers. This latter subject really forms a study in itself, but is sufficiently touched upon here to cover the commoner causes of trouble.

The books chiefly consulted in the preparation of this work are "Marine Propellers," by Mr. S. Barnaby, and "Ship Form, Resistance, and Screw Propulsion," by Mr. G. S. Baker; and the author desires to express his acknowledgments.

DOUGLAS H. JACKSON.

London,
   1920.

CONTENTS

CHAPTER I
DESIGN

General considerations—Formulæ—Design by comparison—Design from charts—Limited diameter—Number of blades—Correction of diameter for two or four blades—Blade area to suit thrust—Correction for "shaft horse-power"—Efficiency with large blade area—Correction for wake-coefficient—Complete procedure of design - - 1

CHAPTER II
SCANTLINGS

Formulæ—General considerations—Length of boss—Driving keys—Thickness of blade—Thickness of metal of boss—Studs for fixing loose blades - - - - 12

CHAPTER III
WORKING DRAWINGS

Blade form—Projection—Explanation of method—Set-back blades—Blades of increasing pitch—Finishing the drawings - - - - - - - 22

CHAPTER IV
GENERAL THEORY

Resistance and slip—Possible efficiency—Acceleration of propeller stream—Negative slip—Derivation of formulæ—Cavitation—Tip-speed and clearance—Increasing pitch—Augment of resistance—Efficiency—Strength of blades—Trials and tank tests - - - - - 33

CONTENTS

CHAPTER V

PATTERN-MAKING AND MOULDING

Patterns for normal blades—For set-back blades—For varying pitch—Moulding—Sweeping up for normal blades—For set-back blades—For varying pitch - - - 61

CHAPTER VI

MACHINING AND FINISHING

Marking off—Solid propellers—Loose blades—Checking shape—Measuring the pitch—Machining the boss—Finishing blade surfaces—Balancing - - - - 70

CHAPTER VII

REPAIRING

Burning-on—Electric welding—Autogenous welding—Replacement—Measurement for replacement—Removing the propeller for repair or replacement - - - 78

APPENDIX

Definitions - - - - - - - 84
Tables - - - - - - - 86
Index - - - - - - - 91

LIST OF TABLES

TABLE		PAGE
I.	CALCULATION OF BLADE STRESSES	18
II.	CALCULATION OF STRESSES IN STUDS	20
III.	BLADE-FORM ORDINATES	31
IV.	AREAS OF CIRCLES	86
V.	SQUARES	87
VI.	SQUARE ROOTS	88
VII.	FIFTH POWERS	88
VIII.	CUBE ROOTS	89
IX.	KNOTS, MILES, KILOMETRES	90

LIST OF ILLUSTRATIONS

FIGURE		PAGE
1.	EFFECTIVE THRUST PER SQUARE INCH OF BLADE	9
2.	PROPORTIONS OF DRIVING KEYS	16
3.	UNSUITABLE BLADE FORMS	23
4.	METHOD OF PROJECTION MODIFIED TO GIVE AN APPROXIMATE DEVELOPED SHAPE OF BLADE	27
5.	CHANGES IN VELOCITY OF PROPELLER STREAM	35
6.	DIAGRAMMATIC REPRESENTATION OF PROPELLER STREAM	39
7.	EFFECTIVE SECTIONAL AREA OF PROPELLER STREAM	40
8.	EFFECT OF BLADE-THICKNESS AND SURFACE FRICTION	45
9.	SINGLE SCREW IN ORDINARY STERN FRAME	48
10.	EFFECT OF OVERLAPPING OF PROPELLER CIRCLES	49
11.	WAKE COEFFICIENTS	53
12.	POSSIBLE EFFICIENCY	54
13.	METHOD OF CHECKING BLADE-STRESSES	59
14.	PATTERN FOR NORMAL BLADE	62
15.	CONSTRUCTION OF PATTERN FOR SET-BACK BLADE	63
16.	BLANK PREPARED FOR PATTERN OF INCREASING-PITCH BLADE	65
17.	SWEEPING-UP A MOULD	66
18.	MOULDER'S TEMPLATES FOR INCREASING-PITCH PROPELLER	67
19.	DIAGRAMMATIC ARRANGEMENT OF FEEDERS AND RISERS	68
20.	PROPELLER WITH PITCH-PLATES IN POSITION	71
21.	PITCHOMETER	73
22.	TEMPLATES FOR BORE OF PROPELLER BOSS	75
23.	METHOD OF TESTING BALANCE	76
24.	TYPE OF FRACTURE CAPABLE OF REPAIR	79
25.	MEASUREMENT OF PROPELLER FOR REPAIR OR REPLACEMENT	81
26.	METHODS OF WITHDRAWING PROPELLERS FROM SHAFTS	82

LIST OF PLATES

PLATE
- I. CHART FOR PROPELLER DESIGN—"X" VALUES
- II. CHART FOR PROPELLER DESIGN—"Y" AND "Z" VALUES
- III. CHART FOR PROPELLER DESIGN—DIAMETERS
- IV. PROPELLER BLADE SCANTLINGS AND SECTIONS
- V. PROJECTION OF NORMAL TYPE OF PROPELLER BLADE
- VI. PROJECTION OF SET-BACK TYPE OF PROPELLER BLADE
- VII. PROJECTION OF SET-BACK BLADE WITH INCREASING-PITCH

} *at end.*

DETAIL DESIGN OF MARINE SCREW PROPELLERS

CHAPTER I

DESIGN

General considerations—Formulæ—Design by comparison—Design from charts—Limited diameter—Number of blades—Correction of diameter for two or four blades—Blade area to suit thrust—Correction for "shaft horse-power"—Efficiency with large blade area—Correction for wake-coefficient—Complete procedure of design.

General Considerations.—In preparing a new design it is frequently necessary to fix upon approximate dimensions for the diameter and pitch of the propeller at a very early stage, in order that the shaft line may be settled. These preliminary figures have usually to be calculated in short time, with little opportunity for full investigation of conditions, and no opportunity at all for model experiments. Two principal methods of calculation are then available: Firstly, direct comparison with one or more existing propellers of known performance: and, secondly, calculation based on tables or charts derived from systematic experiments. Both of these methods will be described in the following pages.

Certain particulars in connection with the design are usually fixed by considerations independent of the propeller itself. These are, as a rule: the speed of the vessel, the power to be transmitted by one propeller, and the engine revolutions. Of these data, the second and third may be subject to modification to suit the propeller design.

Formulæ.

(1) $D \propto \sqrt[3]{\dfrac{I.H.P.}{R}}$

(2) $D \propto \sqrt[2]{\dfrac{I.H.P.}{S^3}}$

(3) $D \propto \sqrt{\dfrac{I.H.P.}{S}}$

(4) $R \propto \dfrac{S}{D}$

(5) $X = \dfrac{I.H.P. \times R^2}{S^5}$

(6) $D^2 \cdot \sqrt{\delta} = D_1^2 \sqrt{\delta_1}$

(7) $D = K \cdot \sqrt[3]{\dfrac{I.H.P.}{R}}$, where $K = \left(\dfrac{26}{S} + 2\cdot 74\right)$, where

D = diameter of propeller in feet.
I.H.P. = indicated horse-power transmitted through one propeller.
R = revolutions per minute.
S = speed of vessel in knots.
δ = (developed area) ÷ (disc area).

§ 1. **Design by Direct Comparison.**—For cases of close similarity of speed, Formula 1 is suitable. If D_1, $I.H.P._1$, and R_1 be used to denote particulars of an existing propeller, and D_2, $I.H.P._2$, and R_2, those of the new design, then:

$$D_2 = D_1 \sqrt[3]{\dfrac{I.H.P._2}{I.H.P._1} \cdot \dfrac{R_1}{R_2}}.$$

Where the ratio of engine revolutions to speed of vessel is similar in the existing and projected designs, Formula 2 may be used. Then using D_1, D_2, etc., as above:

$$D_2 = D_1 \sqrt{\dfrac{I.H.P._2}{I.H.P._1} \cdot \dfrac{S_1^3}{S_2^3}}.$$

DESIGN

Again, if the relation between speed, power, and revolutions be similar in existing and projected designs, Formula 3 is applicable and may be written thus:

$$D_2 = D_1 \sqrt{\frac{I.H.P._{\cdot 2}}{I.H.P._{\cdot 1}} \cdot \frac{S_1}{S_2}}.$$

In connection with Formula 3 useful information can be obtained by drawing curves in which $\dfrac{D}{\sqrt{\dfrac{I.H.P.}{S}}}$ is plotted against the block-coefficient of the vessel.

It must be emphasized that the value of such curves depends entirely on their use for similar types of hull-form, irrespective of actual block-coefficients. Thus, results obtained from full-bodied cargo ships cannot be grouped with those from flat-bottomed shallow-draft vessels, although the block-coefficients might be similar. No such curve is presented here, for that reason.

It will be observed that the above formulæ do not give any guidance as to pitch or blade area. As regards pitch the usual practice is to proportion "apparent slip" according to speed and revolutions. This is a rather "rough and ready" method, and though it has little basis in theory it answers very well in practice. A very rough rule may be given as follows, where engine revolutions are suitably proportioned to speed of vessels:

$$\text{Percentage of apparent slip} = \frac{S}{2} + 1 \quad \ldots \quad (8)$$

It is generally accepted that the best results are obtained with a pitch ratio between 1·2 and 1·5.

§ 2. **Design from Chart or Formula.**—An empirical formula by which the propeller diameter may be roughly settled is stated as follows:

$$D = K \sqrt[3]{\frac{I.H.P.}{R}}, \text{ where } K = \frac{26}{S} + 2\cdot74 \text{ (Formula 7)}.$$

This is, of course, a variation of No. 1. Diameters obtained by this use of this formula are too great to be practicable for high-speed vessels.

Formulæ 2, 4, and 5, which are discussed together with Nos. 1, 3, and 6 in Chapter IV., were established by Froude. From the results of the Thornycroft experiments Mr. Barnaby prepared a table of coefficients for the determination of propeller dimensions. The charts shown on Plate II. have been prepared, with slight modifications, on the basis of those coefficients. For a fuller treatment of the matter the reader is referred to Mr. Barnaby's book and papers.

The formulæ in question are expressed in the following form:

$$X = \frac{I.H.P. \times R^2}{S^5}.$$

$$Y = \frac{(\text{disc area}) \cdot S^3}{I.H.P.};$$

hence disc area $= \dfrac{Y \cdot I.H.P.}{S^3}$, which is Formula 2 in another shape.

$$Z = \frac{R \cdot D}{S};$$

hence $R = \dfrac{Z \cdot S}{D}$ (Formula 4).

In the charts, Y and Z are plotted against corresponding values of X. The curves are drawn for three-bladed propellers of 30 per cent. developed area, as this type is by far the most common in present-day new construction. The methods of correcting results for modifications in design, together with the relative advantages of two-, three-, and four-bladed propellers, are discussed later in this chapter.

In the most usual cases where speed, power, and revolutions are fixed, the charts will be used in the following manner: The value of X is obtained without calculation from Plate I. by finding the point of intersection of the

DESIGN

speed line with the appropriate revolution line. The height of this point above the base-line of the chart is then traced along to the correct power line. Below this intersection read the value of X.

Example 1.—A numerical example will serve to make this clear: it is required to find the value of X for 5,000 i.h.p., 300 revolutions per minute, 25 knots. Tracing down the 25-knot line, from the scale at the top of the chart, it is found to cut the 300 r.p.m. line about eight divisions above the base-line. If this level be traced along to cut the 5,000 i.h.p. line, the value of X is read off on the scale at the bottom of the chart as 46.

Now, referring to Plate II., the value obtained for X is noted on the scale at the bottom edge of the chart. A series of values of Z for different pitch ratios can then be read off from the chart, with the corresponding efficiencies as shown by the efficiency curve at the top of the chart. Diameters and pitches are calculated from the formula $D = \dfrac{Z \cdot S}{R}$, and the most suitable figure selected.

Example 2.—Taking as an example the figures noted above, the following figures are obtained for an X value of 46:

No.	Z.	Efficiency.	Pitch Ratio.	Diameter.	Pitch.	Apparent Slip.
				Feet.	Feet.	Per Cent.
1	108·5	0·69	1·0	9·03	9·03	6·7
2	102·5	0·695	1·1	8·52	9·37	10·0
3	97·0	0·69	1·2	8·08	9·7	13·0
4	93·0	0·69	1·3	7·75	10·08	16·3
5	89·5	0·68	1·4	7·46	10·44	19·3
6	87·0	0·67	1·5	7·25	10·87	22·4

If no other consideration were involved, the proportions given in lines 2 or 3 might be adopted. Compare these results with those given by Formulæ 7 and 8:

$$D = \left(\frac{26}{25} + 2\cdot74\right)\sqrt[3]{\frac{5000}{300}} = 9\cdot67 \text{ feet, which is far too large.}$$

Slip = 13·5 per cent., and P = 9·75 feet, which agrees well with line 3 above.

In the less usual case where only speed and power are fixed, and it is required to find diameter and pitch of the propeller as well as the most suitable speed of revolution, use is made of both charts on Plate II. The method can best be illustrated by a numerical example.

Example 3.—Assume that a design is required for a propeller to transmit 3,200 i.h.p., the speed of vessel being 25 knots. The engine revolutions are to be arranged to suit a pitch ratio of about 1·5. Proceed as follows:

On the lower chart of Plate II. locate the intersection of the 1·5 pitch-ratio line with the efficiency line of highest value. Trace this point up to the corresponding efficiency line in the upper chart. Repeat the process for 1·4 and 1·6 pitch ratios and tabulate the results thus:

No.	X.	Y.	Z.	Pitch Ratio.	Efficiency.	Diameter.	Pitch.	R.P.M.	Apparent Slip.
									Per Cent.
1	19·2	238	82	1·4	0·695	7·9′	11′	260	11·5
2	15·9	250	77	1·5	0·695	8·1′	12·1′	238	12·0
3	13·6	264	73	1·6	0·695	8·3′	13·3′	220	13·4

Of the results thus obtained, the set giving the most suitable engine speed would be chosen if no other considerations were involved. Note: the chart given on Plate III. shows graphically the relations between I.H.P., D, S, and Y.

§ 3. Limited Diameter.—In many cases of modern practice, the permissible diameter of the propeller is severely limited, either by depth of water or by peripheral speed. The limit of peripheral speed, with good immersion and ample hull clearance, may be taken as 280 feet

DESIGN

per second (17,000 feet per minute), though propellers have been run at 300 feet per second peripheral speed. It is desirable, however, to keep well below that figure.

If the maximum diameter, consistent with these conditions, be less than that calculated from the charts, Formula 6 may be applied, as in the following example:

Example 4.—Assume that for the propeller considered in Example 2, No. 3, the maximum diameter allowable is 7 feet. Then $8 \cdot 08^2 \cdot \sqrt{0 \cdot 3} = 7^2 \cdot \sqrt{\delta_1}$, where δ has a maximum value of 0·55.

In this case $\delta_1 = 0 \cdot 53$. So that the final proportions for the propeller would be: diameter 7 feet, pitch 9′ 8½″, developed area 39 square feet, pitch ratio 1·385.

§ 4. **Number of Blades.**—The tendency of modern practice is to make the three-bladed propeller practically the standard type. At the same time, two-bladed and four-bladed propellers have certain advantages. The two-bladed propeller makes the simplest and cheapest casting, and is easy to balance correctly. It can be blocked in a vertical position to lie behind a stern-post, with a view to minimizing resistance when the vessel is being towed or is proceeding under sail. On the other hand, for vessels of moderately high speed a high value of δ involves great breadth of blades, with consequent increased liability to deformation.

The four-bladed propeller is particularly suitable for vessels of shallow draft, where only small immersion can be obtained. In such cases the constantly varying thrust on each blade due to its changing immersion may produce a serious irregularity of torque with accompanying vibration. Such irregularity is, of course, reduced in some measure by increasing the number of blades. The same remarks apply with even greater force to the case of vessels making extended voyages "in ballast"—*i.e.*, with the propeller circle only partly immersed. That practice is, of course, avoided as far as possible in these days. The four-bladed propeller is possibly from 1 to

8 MARINE SCREW PROPELLERS

2 per cent. less efficient than the three-bladed type. This is a maximum discrepancy, and is offset by the greater root-thickness necessary for a three-bladed screw working at the same thrust; so that in practice there is little or nothing to choose in efficiency between the two types.

§ 5. **Correction of Diameter for Two or Four Blades.**—To correct the propeller diameter found according to § 2 to suit the case of two-bladed or four-bladed propellers, proceed as follows:

$$\text{diameter (two blades)} = D\sqrt[4]{\tfrac{3}{2}} = 1\cdot067\ D$$
$$\text{diameter (four blades)} = D\sqrt[4]{\tfrac{3}{4}} = 0\cdot929\ D$$

where D is the diameter found from chart.

This is an application of Formula 6, and gives diameters for propellers of which each blade has a developed area of 10 per cent. of the disc area.

§ 6. **Correction of Blade Area for Thrust.**—In order to avoid cavitation (See Chapter IV., § 6) it is necessary to provide a certain minimum blade area according to the effective thrust. This effective thrust may be reckoned at $\dfrac{\text{I.H.P.} \times 33{,}000 \times 0\cdot5}{S \times 101\cdot3}$, where 0·5 is taken as the over-all efficiency of the propelling installation. This efficiency may in certain cases rise as high as 0·65, but 0·5 is a safe average value.

The maximum allowable intensity of thrust in pounds per square inch of *projected* blade area $= \left(10\cdot85 + \dfrac{3h}{8}\right)$, where h is the depth of immersion of the uppermost point of the propeller circle in feet (Barnaby's formula). This is suitable for well-balanced reciprocating engines. For four-cylinder internal combustion engines (4-stroke cycle), where the torque is more uneven, the intensity of thrust should not exceed 8 or 9 lb. per square inch, while for turbine-driven propellers it may be as much as 13 or 14 lb. per square inch. Figure 1 shows the average pressures allowed in practice plotted against the speed of the vessel. It will be noted that the blade area may be made as much as 80 per cent. or even more of the disc area, to avoid cavitation, although no extra thrust is obtained by increasing the area beyond 55 per cent. of the disc area.

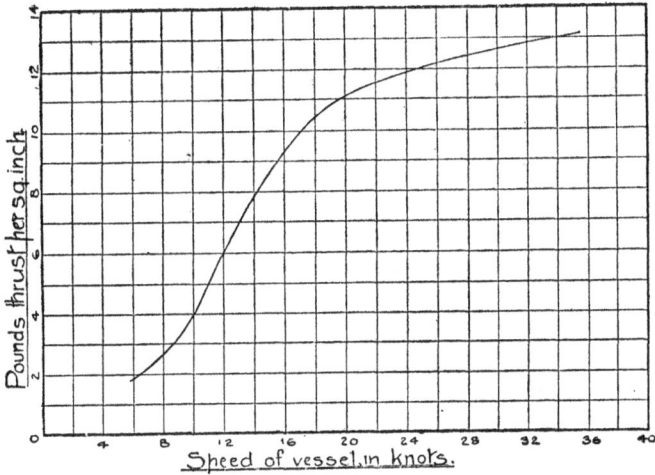

FIG. 1.—EFFECTIVE THRUST PER SQUARE INCH OF BLADE.

§ 7. **Correction for "Shaft Horse-Power."**—The charts on Plate II. and all of the above formulæ and constants are based on the indicated horse-power of reciprocating steam engines, with an average mechanical efficiency of 0·8 to 0·85. Therefore if it is required to calculate from the shaft horse-power of turbines or the brake horse-power of internal combustion engines a correction should be made thus:

$$\text{I.H.P.} = \frac{\text{S.H.P.}}{0\cdot 9} \text{ or } \frac{\text{B.H.P.}}{0\cdot 85}$$

The discrepancy which will be noted in this formula, as regards the correction for S.H.P., is an allowance for the more even turning moment obtained with turbine machinery.

§ 8. Correction of Efficiency for Blade Area.

—The efficiencies noted on the chart are approximately correct for the proportions stated—*i.e.*, for propellers having a total developed area of 30 per cent. of the disc area. If the blade area be increased beyond this proportion, for reasons noted in § 6, the efficiency falls off in the following manner:

Developed area / Disc area.	Deduction from Efficiency shown on Chart, Plate II.			
	Pitch Ratio, 0·8.	1·0.	1·2.	1·5.
Per Cent.				
40	0·01	0·005	—	—
50	0·025	0·01	0·005	—
60	0·045	0·02	0·01	0·005
70	0·075	0·04	0·02	0·01
80	0·11	0·055	0·03	0·02

It should be clearly understood that statements of propeller efficiency must always be regarded as approximate only, since the possible efficiency varies very considerably according to the nature of the blade surface. The curve, Figure 12, shows efficiencies obtained in the course of careful experiments by Baker, Froude, and others. It is in the highest degree unlikely that such efficiencies would ever be realized in practice.

§ 9. Correction for Wake-Coefficient.

—The charts on Plate II. are drawn for a w value of 0·9, which may be accepted as correct for general use. If the value of w in a particular design is known with reasonable certainty to be different from the above, correction should be made as follows:

$$S = \frac{(\text{speed of vessel}) \cdot w}{0 \cdot 9}.$$

The value of S thus obtained is used in place of the actual speed of the vessel, for X, Y, and Z.

DESIGN

§ 10. **Complete Procedure.**—The complete procedure of design may be set out as follows; it will, of course, be understood that some of the corrections indicated may be unnecessary in a given case:

(a) Decide number of blades and limiting diameter.

(b) Correct S.H.P. to I.H.P.

(c) Correct speed for given wake-coefficient.

(d) Find value of X from chart, Plate I.

(e) Find corresponding values of Y and Z and tabulate as in Example 2.

(f) Correct diameter for two (four) blades.

(g) Correct for limiting diameter and check tip-speed.

(h) Increase blade area as necessary for thrust.

(j) Check efficiency to suit modification.

NOTE FOR PLATE III.

Method of using Plate III.: Take the value of "Y" on the scale at the top edge of the chart, trace it down to the intersection with the appropriate I.H.P. line. The value of Y × I.H.P. is then read on the scale at the left-hand side. Trace this value horizontally to intersect the appropriate speed line, and read the diameter on the scale at the lower edge of the chart.

CHAPTER II

SCANTLINGS

Formulæ—General considerations—Length of boss—Driving keys—Thickness of blade—Thickness of metal of boss—Studs for fixing loose blades.

Formulæ—

(1) $T = \frac{1}{2}$ inch per foot of diameter for cast iron, $\frac{3}{8}$ inch for gunmetal, $\frac{5}{16}$ inch for cast steel or bronze.

(2) $T_1 = \sqrt{\dfrac{d^3}{n \times b} \times K} + C$ (Seaton's formula).

(3) $T^2 = \dfrac{N \times I.H.P. \times (D - d_1)}{B_1}\left(\dfrac{d_1}{P \times S} + \dfrac{20}{R.D.}\right)$
(Barnaby's formula).

(4) $M = \dfrac{B \times T^2 \times R \times P}{IHP \times (D - d_1)}$ (Admiralty formula).

(5) $T = J\sqrt{\dfrac{IHP.D}{S.B.}}$.

T = thickness of blade at surface of boss in inches.
T_1 = thickness of blade at centre of shaft in inches; see Plate IV. (A).
d = diameter of tail-end shaft in inches.
d_1 = diameter of boss in feet.
D = diameter of propeller in feet.
n = number of blades in one propeller.
b = breadth of blade at root in inches, measured parallel to shaft; see Plate IV. (B).
B = breadth of blade at root in inches, measured as in Plate IV. (C).
B_1 = breadth of blade at root in inches, measured as in Plate IV. (D).

SCANTLINGS

R = revolutions per minute.
P = pitch of screw in feet.
IHP = indicated horse-power transmitted *through one blade*.
S = speed of vessel in knots.
C = ¼ inch for gunmetal, bronze, or cast steel; ½ inch for cast iron.
K = 2 for gunmetal; 1·5 for cast steel or hard bronze; 4 for cast iron.
N = 4 for gunmetal; 2 for cast steel or hard bronze; 6 for cast iron.
M = 90 (slow speed) to 130 (high speed) for hard bronze.
J = ·8 to ·9 for cast steel or hard bronze; 2 to 2·25 for cast iron.

§ 1. **General Considerations.**—After settling upon the leading particulars of a propeller as described in the previous chapter, two important decisions have to be made. They are (*a*) the material of which the screw is to be made, and (*b*) whether the screw is to be cast solid, or to have loose blades.

Now the first point is largely a matter of custom, such as cast iron propellers for trawlers, cast steel or cast iron for tramp steamers, and bronze for fast liners, cross-channel steamers, etc. Of course, special circumstances frequently determine the material to be used, but speaking generally, the prime cost of propellers for tramp steamers is kept as low as possible. This is, more often than not, false economy. A well-finished bronze propeller has far smaller frictional losses than a cast iron or cast steel screw, and generally keeps its smooth surface very much better. The cast steel blade is particularly bad as a rule from the friction point of view, and is seldom of the designed shape and pitch.

The second point (solid or loose blades) is also largely a matter of custom. It is, however, now increasingly common to cast all propellers solid up to about twelve feet in diameter, except in special cases where frequent damage to blades (*e.g.*, from ice) is likely to occur. Propellers of more than about 12 feet diameter are nearly always of the loose-blade type.

In connection with loose-bladed screws, a not uncommon practice is to use a cast steel boss, with bronze blades. The wisdom of this policy is for several reasons doubtful. Galvanic action between the boss and the blades is positively invited; moreover, a cast steel boss has an annoying way of " growing on " to the tail-end shaft in such a way that little short of an earthquake is needed to withdraw it.

Forged steel blades, such as were used in some of the early torpedo craft, are obsolete, owing to the great cost of their manufacture and their liability to corrosion.

§ 2. **Length of Boss.**—The length of boss, to suit, on the one hand, the blade, and, on the other, the driving key or keys, should be from two to three times the diameter of the tail-end shaft. The boss is almost invariably bored taper to receive the shaft. A few words may be said concerning this taper: (*a*) Its purpose is *not* to transmit the torque from the shaft to the propeller; keys are provided for that purpose. (*b*) Its purpose is to ensure a proper bearing of the propeller boss on the shaft in spite of small discrepancies between the diameter of the shaft and the bore of the boss. (*c*) The propeller must be capable of being removed from the shaft, without damage to either.

The actual taper of the bore is commonly made 1 inch of diameter in 12 inches of length. One inch in 16 inches is also common, but is unnecessarily fine; 1 inch in 10 inches is also used by some first-class builders, and is probably the best figure to adopt. In passing, it may be noted that the coupling flange at the forward end of the tail-end shaft should be made sufficiently strong to withstand the strains put upon it in removing the propeller.

§ 3. **Driving Keys.**—The usual mercantile practice is

SCANTLINGS

to fit one key only. This makes for cheapness and simplicity of fitting. In naval work, however, and in cases where the size of shaft is kept down to the lowest safe limit, two keys should be fitted. They should be arranged opposite to one another—*i.e.*, 180 degrees apart.

Keys may be proportioned as follows:

Crushing load on side of key in shaft: maximum stress: 22,000 lb. per square inch.

Crushing load on side of key in propeller: maximum stress: 18,000 lb. per square inch.

Maximum shear stress on key: 8,000 lb. per square inch.

Where two keys are fitted, each key should be made suitable for two-thirds of the total torque. The chart (Fig. 2) gives sizes of keys based on the above figure, and the assumption that the effective length of the key (*i.e.*, the length actually bearing at the sides on both shaft and boss) is not less than 1·5 times the diameter of the shaft.

§ 4. **Thickness of Blade.**—The thickness of blades at the circumference may be made equal to

$(·03 D + ·10)$ inches, for gunmetal or bronze.
$(·04 D + ·15)$,, ,, cast steel.
$(·05 D + ·20)$,, ,, ,, iron.

The thickness at the root is then provisionally settled by means of one of the formulæ given at the head of this chapter. These two thicknesses are then set down to scale, on a drawing (see Plate VI.), and joined by a straight line.

If, now, the screw under design is one of normal proportions—that is to say, not of extremely wide-bladed type, and not running at very high revolutions—and if, above all, the material be cast iron or cast steel, the blade thickness may be counted as settled. If, however, the design of the propeller involves consideration of some special features, such as great blade area for its diameter,

16 MARINE SCREW PROPELLERS

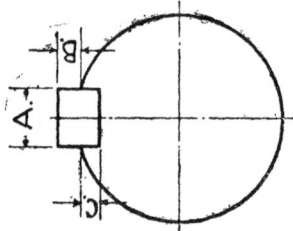

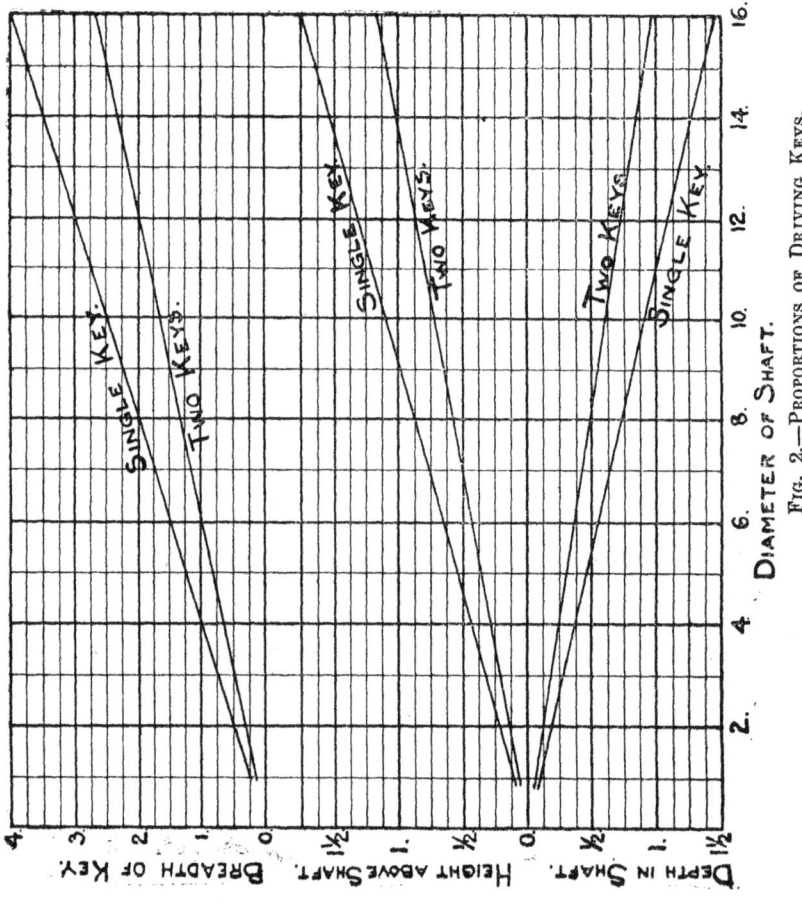

Fig. 2.—Proportions of Driving Keys.

SCANTLINGS

high revolutions, etc., then the stresses on the blade should be checked. If, again, the blades are to be of gunmetal or bronze, then obviously the thickness should be kept down as far as is consistent with strength to avoid expense.

The method suggested for checking stresses is as follows: Referring again to Plate VI., the distance $a\,b$ is divided into four equal parts (note that this has been done in setting out the blade shape; see Chapters I. and III.). The developed area of each strip of blade is then run out by planimeter, and set down in tabular form (see Table I., p. 18, line B).

The figures given in this table are worked out for the propeller drawn on Plate VI., which has the following particulars: Diameter 12', pitch 18', total I.H.P. 3,200, R.P.M. 170, speed 25 K., total thrust 20,850 lbs.

Line A gives the area of stress-section, taken as $\frac{2}{3}$ (breadth) × (thickness). The volume of the strip is given by $\frac{2}{3}$ (line B) × (line C), which, multiplied by the appropriate figure, according to the material of the blade, gives the value for line D. Line E is obtained by—

$$\frac{(\text{line D})}{32\cdot 2} \times \left(\frac{R \times 2\pi}{60}\right)^2 \times (\text{radius from centre of shaft to C.G. of strip, in feet}).$$

The value of F for column 1 is equal to the value of E for column 1;

F for column 2 equals the sum of the values of E in columns 1 and 2;

F for column 3 equals the sum of the values of E in columns 1, 2, and 3; and

F for column 4 equals the sum of the values of E in columns 1, 2, 3, and 4.

Line G is obtained thus:

$$\frac{(\text{IHP per blade}) \times 33{,}000 \times \cdot 5 \times (\text{line B})}{101\cdot 3 \times (\text{speed in knots}) \times (\text{dev. area of 1 blade})}.$$

TABLE I.

Line.	Item.	1	2	3	4
A	Area of stress section, square inches	39·6	65·3	76·6	71·4
B	Area of strip, square inches	456	606	579	460
C	Thickness at C.G. of strip, inches	$1\frac{1}{16}$	$1\frac{3}{16}$	$2\frac{1}{2}$	$3\frac{5}{16}$
D	Weight of strip, lb.	103	233	307	323
E	Centrifugal force, lb.	5,300	10,020	9,670	5,765
F	Total centrifugal force, lb.	5,300	15,320	24,990	30,755
G	Thrust on strip, lb.	1,508	2,005	1,915	1,522
H_1	Bending moment due to thrust, lb. inches	6,600	—	—	—
H_2	Bending moment due to thrust, lb. inches	26,950	12,400	—	—
H_3	Bending moment due to thrust, lb. inches	47,200	39,350	13,160	—
H_4	Bending moment due to thrust, lb. inches	67,500	66,400	39,000	11,040
J_1	Bending moment due to centrifugal force, lb. inches	2,320	—	—	—
J_2	Bending moment due to centrifugal force, lb. inches	11,920	8,150	—	—
J_3	Bending moment due to centrifugal force, lb. inches	20,560	24,420	8,460	—
J_4	Bending moment due to centrifugal force, lb. inches	29,120	41,400	24,180	4,960
K	Total bending moment, lb. inches	8,920	59,420	153,150	283,600
L	Stress due to bending moment, lb. per square inch	983	2,500	4,070	6,100
M	Stress due to centrifugal force, lb. per square inch	134	234	326	431
N	Total stress, lb. per square inch	1,117	2,734	4,396	6,531
P	Thickness as modified, inches	$1\frac{1}{16}$	$1\frac{11}{16}$	$2\frac{7}{16}$	—

Lines H_1, H_2, H_3, H_4 show the moments of forces given in line G, acting at the centres of area of the strips, about the stress-sections.

SCANTLINGS

Lines J_1, J_2, J_3, J_4 show the moments of forces given in line E acting at points marked "C.G. of strip" about the centres of stress-sections. These figures only appear in the case of "set-back" blades.

Line K gives the sum of lines H and J.

The figures in line L are made equal to:

$$(\text{line K}) \div \frac{(\text{breadth of stress-section}) \times (\text{thickness of stress-section})^2}{9}.$$

Line M equals (line F) ÷ (area of stress-section).

Line N gives the sum of figures in lines L and M.

Line P shows modified thickness of stress-sections to bring line N more approximately to a constant figure throughout the blade.

§ 5. **Thickness of Metal of Boss.**—For solid-cast screws the thickness of metal of the boss must be considered in relation to the thickness at the root of the blade. For steel or bronze propellers the proportions shown in Plate IV. may be relied on to give a good casting; while for cast iron the boss-thicknesses shown may be increased not more than, say, 25 per cent., such increase being accompanied by an increase in the radius of the fillets at the back and front of the blades. It is common practice to enlarge the bore of the boss at the centre, for about one-third of its length. The recess thus formed may be cast or machined out. Reference is made to this point in Chapter III. For set-back blades, the radius of the fillet at the *back* of the blade needs to be substantially increased.

§ 6. **Studs for Fixing Loose Blades.**—The studs for fixing loose blades to the boss are commonly made in odd numbers, arranged with one more on the after side of the blade than on the forward side. The proportions may be *provisionally* settled thus:

(Thickness of blade-flange at pitch circle of studs) = (thickness at root of blade) × ·6.

20 MARINE SCREW PROPELLERS

Diameter of studs = thickness of flange.

Number of studs = 7 (four aft, three forward) as a minimum.

Pitch-circle diameter = 7·5 (thickness of flange).

These rules are given for guidance only. Sizes obtained by them should be checked in the following manner:

Referring again to Plate IV., tabulate thus:

TABLE II.

	1 and 2	3 and 4	5 and 6	7
Sectional area of studs at bottom of threads in square inches	5·10	5·10	5·10	2·55
Arm about centre of pitch-circle in inches	3·5	8·5	6·5	9
Moment of inertia	62·5	368	215	207

From the total moment of inertia approximately obtained by this method, the modulus of the system is calculated in the usual manner. In the example given, $Z_t = 89·5$.

Then the stress due to bending =

$$\frac{\text{bending moment, line K, column 4, Table I.}}{Z_t}.$$

And the stress due to centrifugal force =

$$\frac{\text{centrifugal force, line F, column 4, Table I.}}{\text{total area of studs at bottom of threads}}.$$

The sum of these two stresses gives the maximum stress on the studs, which should be in accordance with usual practice, according to the size of stud and the material used. For steel studs the total stress should never exceed 10,000 lb. per square inch; while studs of a 25-ton bronze should not be stressed above 8,000 lb. per square inch.

Mr. Seaton gives the following rule for the size of studs;

needless to say, it is thoroughly reliable for all normal propellers:

$$a \times N \times r = \frac{T \times L}{K},$$

where a = area of one stud at bottom of thread in square inches.

N = number of studs.

r = radius of pitch circle in inches.

T = *indicated* thrust per blade.

L = ·6 × total length of blade (flange joint to tip) in inches.

$K = \begin{cases} 1{,}700 \text{ for steel studs.} \\ 1{,}400 \text{ for naval brass or bronze studs.} \end{cases}$

CHAPTER III

WORKING DRAWINGS

Blade form—Projection—Explanation of method—Set-back blades—Blades of increasing pitch—Finishing the drawings.

§ 1. Blade Form.—In the previous chapter we have discussed methods of settling the principal dimensions of a screw propeller to fulfil given conditions, together with the scantlings of blades and boss. The last point for the designer is the preparation of working drawings. In this connection the question of "blade-form" needs some consideration. By this expression is meant the actual shape of the developed, or the projected, blade. Much has been written, and many more or less ingenious theories have been formulated, on this subject. Curious schemes have been patented, and equally curious models have been pressed upon the notice of contractors and consulting engineers. Very few of these devices have achieved any outstanding measure of success. A few general conclusions may be mentioned, as being supported by sound theory as well as by practical experience. Thus, the outline of the blade should clearly be an easy and fair curve. "Corners," such as those shown in Fig. 3, at (*a*), and excessively overhanging shapes as indicated at (*b*), Fig. 3, do not serve any really useful purpose, and are peculiarly liable to deformation under water thrust. Such deformation will give rise to vibration and a decided loss of efficiency.

Again, the sharply angled part of the blade near the boss is likely to be somewhat inefficient, especially as it is also the thickest portion. Attempts have been made

WORKING DRAWINGS

to remedy this point by reducing the pitch near the boss. Thus, in overcoming one difficulty another one is encountered, as such difference of pitch gives rise to eddying. Better results may be obtained by preserving a uniform pitch, in conjunction with a fairly large boss.

The "elliptical" blade has for long been regarded as the standard form. Where the area of each blade does not exceed 10 per cent. of the whole disc area, this type has many merits; but where this proportion is exceeded,

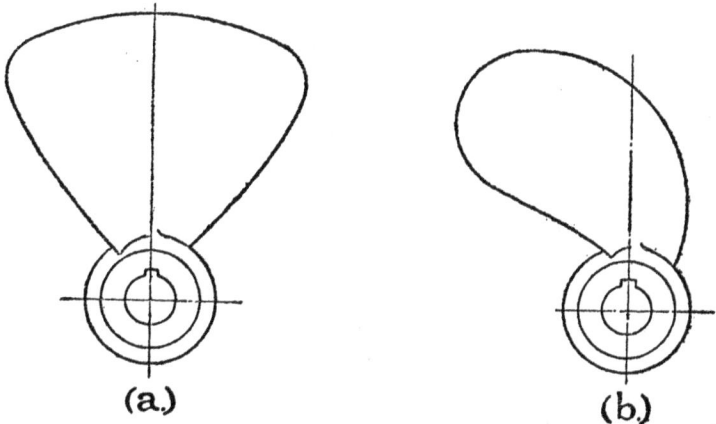

Fig. 3.—Unsuitable Blade Forms.

as is becoming more and more often the case in modern practice, the ellipse is unsuitable owing to the excessive width near the boss. There seems to be no special virtue about the ellipse itself, as a mathematical curve (except for ease of setting out), as it is usually applied to the "developed" blade. Now it is, of course, impossible either to develop the *shape* of a helical blade on to a flat surface, or vice versa. The exact developed area of such a blade may be obtained, together with a developed shape, which will vary according to the graphic construction adopted for such development. This being the case, many different blade forms might be obtained from the same "standard elliptical developed blade."

"Set-back" blades offer no proved increase of efficiency over those in which the blades are normal to the shaft line; while the stresses on the root section are certainly increased. "Set-back" should only be adopted where the hull-form makes its use necessary.

A type of blade suggested by the author may be set down from the table of ordinates given on p 31. This table is arranged to suit the system of projection described below. The values of $\frac{\text{developed area}}{\text{disc area}}$ are based on an average boss diameter of one-fifth of the propeller diameter. This table will be found to give a good width of blade at the root together with an easy and fair outline.

§ 2. **Projection.**—Plate V. illustrates the method of projection; it is drawn for the propeller dealt with in the example on p. 17.

The section of boss and blade is first drawn, from figures obtained in the manner previously described. The developed area is then set out, to give the required area and blade form. Note that the area should be reckoned on that part of the blade which lies outside the circle AB, the distance CD being made equal to one-third of the radius Z. From these two views in conjunction, the corrected section of blade may now be obtained as described in the previous chapter. Such correction may, however, be neglected for screws of normal form, as previously noted.

The line FG is next drawn, parallel to the working surface of the blade, the distance EF being equal to $\frac{\text{pitch of propeller}}{2\pi}$.

The developed width QK is then measured along the line EN, from E, giving the point O. This point, being projected on to the line FG, gives the point M. The distance FM is now marked off from K along the arc KH,

WORKING DRAWINGS

which is drawn about the centre of the propeller L. This is readily done, by means of a flexible steel scale, a flexible batten, or other similar means. H is now one point on the outline of the "projected" blade. Now by projecting from H, parallel to FL, and from P perpendicular to FL, the point P is obtained, in the side elevation of the blade. By repeating this process, the complete side and end elevations of the blade are obtained as shown.

§ 3. **Explanation of Projection.**—Before passing on to the dimensioning, etc., of the drawing, a short explanation of this system of projection may not be out of place. The slope of the blade at any point is determined by the ratio of the pitch, to the circumference of the circle passing through that point. That ratio is equal to:

$$\frac{\text{pitch}}{2\pi} \div (\text{radius of the said circle}).$$

Hence the reason for marking off the distance EF equal to $\frac{\text{pitch}}{2\pi}$. The line EN, therefore, shows the slope of the working face at radius FN from the centre. Now, if the triangle EFN were cut out, and were bent so that the edge FN might lie along the arc KH, the triangle would form a *pitch-plate* (see p. 64), and the edge EN would actually form a helical curve which would fit the working face of the blade at radius FN. It is clear that the ratio of $\frac{\text{EN}}{\text{FN}}$ equals $\frac{\text{EO}}{\text{FM}} = \frac{\text{KQ}}{\text{KH}}$, always remembering that the length KH is measured *round the arc*. Again, assume WKHV to be an element of the projected blade. Then the radial distance between WV and KH is constant, and equal to WK. But the true length of WV is equal to WY, and the *true* length of KH is equal to KQ. So that the *true* area of the element WKHV is equal to the area of WKQY.

Thus we have a construction giving correct projections

and correct areas, and involving no helical construction lines. Exception may be taken to the peculiar shape of the developed area. It is, of course, impossible correctly to develop the shape of a propeller blade on to a flat surface; in spite of which, attempts are still frequently made to fit paper templates to the surface of the blades.

If an approximate developed shape is really required, it may be obtained as follows: Referring to Fig. 4, set out a point X, such that the distance LX is about one-tenth of the diameter of the screw. With centre X and radius XK describe the arc KU. With the flexible scale or batten, mark off the distance KU, measured round the arc KU, equal to EO and to KQ. Repeating the process, an approximate developed shape is obtained, which is also of exactly correct area.

§ 4. **Set-back Blades.**—In Plate VI. is shown the method of projecting a screw with "set-back" blades. It will be observed that the construction is precisely the same in principle as that described above. In this case, however, the "developed area" is drawn about a centre line ab, equal in length to FG. The length F_1E_1 is equal to FE (which equals $\frac{\text{pitch}}{2\pi}$), as are also all corresponding lengths, such as F_2E_2. The arcs c, d, e, etc., are drawn with E as centre. Now proceed as before. Mark off the developed width QK along the line E_1N_1 from E_1, giving the point O. This point projected on to the line Gf (perpendicular to FL) gives the point M. The distance fM is now marked off from g along the arc gH which is drawn about the centre of the propeller L. H is now one point on the outline of the "projected" blade. Project from H parallel to FL, and from O perpendicular to FL, to obtain P, which is the point corresponding to H in the side elevation of the blade. Repeating the process, the complete projections of the blade are obtained.

WORKING DRAWINGS

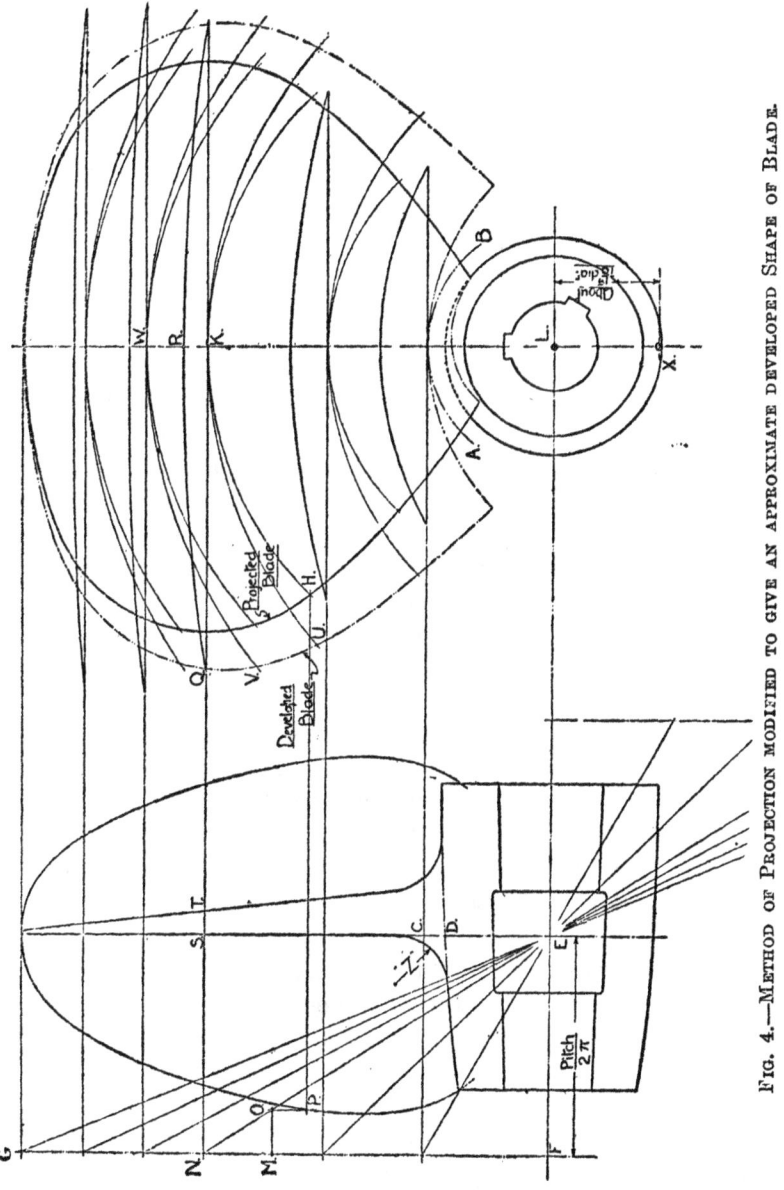

Fig. 4.—Method of Projection modified to give an approximate developed Shape of Blade.

§ 5. Blades of Increasing Pitch.

—The projection of a screw propeller blade having varying pitch is interesting as a piece of drawing, the only difficulty being its involved character.

In Plate VII. is shown the projection of a "set-back" blade in which the pitch varies from root to tip of the blade, as well as from the leading to the trailing edge. It should be distinctly understood that this type of blade is *not* recommended. Variation of pitch from root to tip gives rise to eddying: while variation from leading to trailing edge, though excellent in theory, is seldom of any positive advantage in practice, owing to the impossibility of calculating the requisite amount of such variation. The drawing is, however, given in order to show that such a blade is quite capable of correct projection.

After setting down the developed area, the pitch at tip, root, and leading and trailing edges, is noted on the drawing. From these figures, "curves of pitch" are prepared. On this point, no very precise instructions can be given, as the fairing of the curves depends entirely upon the designer. Two lines such as 1–3–5 and 1–4–6 should be drawn, bisecting the half-breadth of the developed area; and pitch curves on these lines should be faired with the pitch curves on lines 3–4, 5–6, etc. The pitch values obtained from these curves should now be noted at the various points on the developed area.

A blade-slope diagram is next drawn from the figures thus set down. The most convenient way of doing this is to set down a line 8–9 to a convenient scale, equal to (diameter $\times \pi$). By drawing the series of parallel lines 7–9, 10–11, 12–13, etc., we have the corresponding lengths for all stress-sections of the blade. The desired pitch at any point may now be set down along one of the lines 11–14, 13–15, etc., according to the position of the point, and the blade-slope at that point obtained. From these slopes, blade-sections are built up with straight lines,

WORKING DRAWINGS

Referring to the one marked 16–17 (27–28 in developed area), the line 23–24–25 is drawn, equal in length to 18–19–20, on the developed area, and with a slope corresponding to the pitch value of 18′ 0″. From the point 25, the line 25–26 is drawn, equal in length to 20–21, and having a slope corresponding to the pitch of 17′ 0″. Repeating this construction we obtain the lines 26–17, 23–22, 22–16. A fair curve is now drawn to touch these lines at the points 16, 27, 24, 28, 17. The curve is omitted for clearness in Plate VII., as it very nearly coincides with the straight lines. This curve gives the contour of the blade at the section 27–28, and is transferred by means of tracing paper to the side-view of the blade (see line 29–30–31). Note that it is not necessary to set down the complete curve in this view, but merely the points corresponding to the edges of the blade. From these points may now be projected 32 and 33. The distance 34–32 is now transferred to the arc 35–36, giving a point on the projected blade. Further projection gives the corresponding point in the side elevation.

§ 6. **Finishing the Drawings.**—Having prepared correct drawings of the propeller, it becomes important to dimension them in such a way that the pattern-maker or the moulder may know what is required. This is not difficult. In any case of doubt, ten minutes' conversation with an intelligent workman should make all points clear. For blades of constant pitch, the system of dimensions illustrated in Plate V. is quite satisfactory. Blades of varying pitch require rather more elaborate treatment. Probably the best method is to dimension each stress-section with reference to a pure helical surface, as shown in Plate VII. The moulder can then sweep up a helical surface and build up on it the variable pitch surface required. This matter is further discussed in Chapter V.

The shape of stress-sections has been the subject of much experiment and conjecture. Plate IV. shows some types at (*e*) and (*f*) that have been tested, though with no great success. The section shown at (*g*) is commonly adopted, where the working face is flat, and the back is drawn as a circular arc, keeping, of course, a reasonable minimum thickness at the edges.

The contour of the boss and of the cone (if such be fitted) should be fully dimensioned and not "left to the moulder." Where the bore of the boss is recessed, as shown at D in Plate IV., it must be remembered that the effective length of the keys is diminished by that recess. It may be noted here that the founders generally prefer, especially in bronze propellers, that the screw should be cast without a recess in the bore, such recess being afterwards machined out if desired. In any case the axial length of the recess should not exceed about one-third of the length of the boss. See also Chapter II.

A recess should be provided at the forward end of the boss to receive a rubber ring for jointing against the end of the tail-end shaft liner. This recess should not be made too shallow. A ring of $\frac{3}{8}$ inch free thickness can hardly be compressed more than $\frac{1}{8}$ inch. Neglect of this point may mean that the propeller does not bed properly on the tapered end of the shaft. In certain classes of naval work, where no liners are fitted to tail-end shafts, a gland is provided as indicated in Plate V.

Where a cone is fitted at the after end of the boss, it may be attached in several ways. For moderate-sized screws, where the screw-thread on the tail-end shaft is not more than, say, 8 inches diameter, probably the most satisfactory method is to form the cone integral with the nut; see Plate V.

For loose-bladed propellers, Plate VI. shows one form of boss. It will be noticed that the blade flanges are made to form a fair surface with the boss, and that the studs have solid-ended nuts, made of the same material as the blades, to avoid electrolytic action. Both of these latter points are desirable, though perhaps not absolutely necessary. The fairing of the flanges with the boss is frequently neglected altogether, which certainly simplifies the design of boss and flange, though at the expense of efficiency.

The surface of the boss is sometimes faired over in way of the nuts, with cement. This, of course, serves both to provide a smooth surface and to lock the nuts. It is, however, open to the objections that the cement usually breaks away, thus destroying the fair surface, while sufficient remains to make it a matter of extreme difficulty to remove the nuts when desired. A mechanical locking device for the nuts is shown in Plate VI. If such a device be adopted, all parts should be of hard bronze.

It is often the practice to elongate the holes in the flanges of loose blades, so that the blades may be turned through a small angle, to adjust the pitch of the screw. When this is done, a clear and unmistakable system of marking should be adopted, to denote the designed position of the blades. A uniformly pitched blade, twisted in the manner described, becomes of varying pitch from root to tip.

TABLE III.—Blade-Form Ordinates.

Ordinate.	Area of One Blade (developed) as a Fraction of (Disc Area − Boss Area).											
	0.10	0.12	0.14	0.16	0.18	0.20	0.22	0.24	0.26	0.28	0.30	
A	0.147	0.182	0.218	0.253	0.288	0.323	0.359	0.394	0.429	0.465	0.500	Developed breadth of blade expressed as a fraction of propeller diameter.
B	0.184	0.226	0.267	0.309	0.350	0.392	0.434	0.475	0.517	0.558	0.600	
C	0.219	0.265	0.311	0.357	0.403	0.449	0.496	0.542	0.588	0.634	0.680	
D	0.227	0.271	0.316	0.361	0.405	0.450	0.494	0.538	0.582	0.626	0.670	
E	0.198	0.230	0.262	0.293	0.325	0.357	0.389	0.420	0.452	0.484	0.516	
F	0.139	0.159	0.179	0.199	0.219	0.239	0.260	0.280	0.300	0.320	0.340	

The method of using the above table is as follows: The centre line of the blade, from boss-surface to tip-circle, is divided into four equal parts, and the outermost of those parts is further subdivided in the manner shown in Plate V. Positions are thus obtained for six ordinates. The developed breadth of blade at the surface of the boss is the ordinate F in the table, while the outermost ordinate (that nearest the tip of the blade, and figured 2′ 5⅜″ in Plate V.) is marked A in the table. The blade areas given are based on a boss diameter of one-fifth of the propeller diameter: they are sufficiently accurate for all ordinary purposes.

Example.—A three-bladed propeller 10 feet diameter, developed area 60 per cent. of (disc area minus boss area).

Area of each blade = ·20 of (disc area − boss area).

Breadths of blade are: 3·23′, 3·92′, 4·49′, 4·5′, 3·57′, 2·39′.

CHAPTER IV

GENERAL THEORY

Resistance and slip—Possible efficiency—Acceleration of propeller stream—Negative slip—Derivation of formulæ—Cavitation—Tip-speed and clearance—Increasing pitch—Augment of resistance—Efficiency—Strength of blades—Trials and tank tests.

THE following chapter is intended to explain, on simple lines, the broad principles underlying screw propulsion, the principal theories based thereon, and the methods by which these theories may be put to practical use. In order to simplify the matter as far as possible, it has been thought advisable to frame some statements in a way which is not precisely accurate, but which approaches accuracy sufficiently nearly for practical purposes.

§ 1. **Resistance and Slip.**—If a ship, without propeller or other means of self-propulsion, be moved through water at a given speed, a certain propelling force has to be applied which shall be equal to the resistance of the ship to motion at that speed. This resistance to motion is a definite quantity depending upon a number of factors; the principal ones being skin friction and wave making.

(i.) *Frictionless Ship.*—Let us imagine a ship which has *no* skin friction, then we find that the work done in propelling the ship is wholly spent in producing waves—in other words, in moving the surrounding water in a transverse direction. This means that the water has no velocity imparted to it in the direction of motion of the ship.

Now, if this propelling force is to be applied by means of the surrounding water, it is evident that some of that water (in other words, the stream from the propeller) must be pushed backwards by some means, and must have an absolute velocity opposite in sense to that of the ship.

In the discussion that follows let R = resistance of ship in pounds.

W = weight of water pushed back, per second, in pounds.
f = backward acceleration given to the water in feet per sec. per sec.
g = acceleration due to gravity = 32·2 feet per sec. per sec.
P_L = pitch at leading edge of propeller.
P_T = pitch at trailing edge of propeller.
P = mean pitch of propeller.
V = velocity of ship in feet per second.
N = revolutions per second of propeller.

It is evident that the ship-resistance and the weight and acceleration of water handled by the propeller must be related by the well-known dynamical equation:

$$R = \frac{W}{g} f.$$

Note that the backward acceleration, f, is measured with relation to still water.

Suppose the action above described takes place for a space of one second; then the ship has a forward velocity of V feet per second, and the propeller-stream has a backward velocity of f feet per second. That is, the total velocity of the propeller-stream, relative to the ship, is (V + f) feet per second.

Now let us consider the changes of velocity in the propeller stream. Referring to Fig. 5A we may assume the propeller to be revolving without forward motion, and the surrounding water to be moving past the propeller in a backward direction.

Now, if we assume that the whole change of velocity takes place while the water passes the blades of the

GENERAL THEORY

propeller, it is evident that, to avoid shock, the pitch of the propeller must be less at the leading edge than at the trailing edge—that is,

$$P_L = \frac{V}{N},$$
$$P_T = \frac{V+f}{N},$$
$$P = \frac{2V+f}{2N}.$$

Working out the "real slip" of this hypothetical propeller, with reference to the trailing pitch, we find it to be $\frac{f}{V+f}$. It is obvious that there must always be

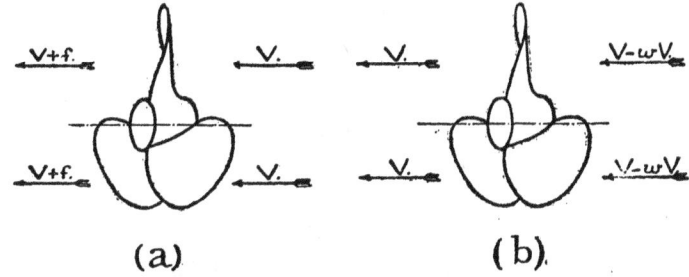

Fig. 5.—Changes in Velocity of Propeller Stream.

this positive slip, under the ideal conditions we have assumed.

(ii.) *Waveless Ship.*—Now, consider another case, that of a ship which forms no waves during its motion through the water. Clearly, the whole of the work done in propelling such a ship would be spent in overcoming skin friction—that is, in imparting a forward velocity to the surrounding water. That forward velocity may be expressed as a function of the velocity of the ship, and may be written as kV.

The surrounding water will then have a kinetic energy equal to $(kV)^2 \times \frac{W}{2g}$.

Now, if we apply from the ship itself a force to neutralize this kinetic energy—*i.e.*, to leave the water in a state of absolute rest—then the ship will be propelled *without* any backward motion of the propeller stream. This looks at first sight like "something for nothing." A little reflection, however, will make the matter clear to the reader.

Fig. 5B shows the changes of velocity of the propeller stream as before—*i.e.*, the propeller is considered as revolving without forward motion, while the surrounding water moves past it in a backward direction. Then, making the same assumption as before—viz., that the whole change of velocity in the propeller stream occurs while the water is passing the blades of the propeller—we find that the pitch at the leading edge should be

$$P_L = \frac{V - kV}{N},$$

and that at the trailing edge should be

$$P_T = \frac{V}{N}.$$

Again, working out the *real slip* of this hypothetical propeller relative to the trailing edge, we find

$$\text{real slip} = \frac{V - V}{V} = \text{zero}.$$

Now the actual state of affairs, when a ship moves through water, is somewhere between these two imaginary conditions of the "frictionless ship" and the "waveless ship." It is evident, therefore, that the stream from the propeller must *always* have an absolute backward velocity. This means that if the propeller fulfilled the condition assumed above (*i.e.*, that the whole change of velocity of the propeller stream occurs in its passage past the propeller), the propeller would always work with positive slip.

GENERAL THEORY

§ 2. Possible Efficiency.—Under the conditions assumed in § 1, section (i.), the work done per second in propelling the ship is equal to RV foot pounds; while that done in producing the backward velocity of the propeller stream is $\dfrac{W}{2g}f^2$.

$$\text{But } R = \dfrac{W}{g}f,$$

$$i.e., f = \dfrac{Rg}{W}, \text{ and } \dfrac{W}{2g}f^2 = \dfrac{R^2 g}{2W},$$

so that useful work = RV,

$$\text{waste work} = \dfrac{R^2 g}{2W},$$

$$\text{total work} = \dfrac{R}{2W}(2WV + Rg),$$

and efficiency $= \dfrac{2VW}{2VW + Rg}$ which can only equal unity when W (the only quantity capable of adjustment) is infinitely great. On the other hand, in the assumed conditions of § 1, section (ii.), since the work done by the ship in accelerating the surrounding water is equal to the work done by the propeller in bringing that water to rest again, the efficiency must equal unity.

Obviously, then, since the real condition lies between those considered in sections (i.) and (ii.) of § 1, advantage should be taken as far as possible of the forward motion of the surrounding water caused by skin friction.

It should now be clear that the possible efficiency of the propeller is definitely limited by the nature of the hull resistance of the ship. The greater the proportion of frictional resistance to the total resistance, the greater will be the possible efficiency of the propeller. That is to say, with a full-bodied vessel the propeller efficiency can never be as high as with one of fine and easy lines.

§ 3. Acceleration of Propeller Stream.—In the foregoing paragraphs it is assumed that the whole change of velocity of the propeller stream occurs whilst the water passes the blades of the propeller. This is not actually the case. Using the same symbols as before, and writing also V_1 as the backward velocity of propeller stream relative to the ship, at the centre of the propeller; then thrust of propeller $= R = \dfrac{W}{g} f$. Work wasted per second on propeller stream $= RV_1 = \dfrac{W}{2g} f^2$,

$$i.e., \dfrac{W}{g} f V_1 = \dfrac{W}{2g} f^2,$$

$i.e., V_1 = \dfrac{f}{2}$, which is to say that the backward velocity of the propeller stream, at the propeller, is one-half of its final velocity.

It was found experimentally by Taylor that there is a reduced pressure of water in front of the propeller, extending as far as one and a half times the propeller diameter. This reduction of pressure is the principal cause of the backward velocity at the propeller. The actual distance through which the propeller receives its change of velocity may then be perhaps three times the propeller diameter.

Consider now a propeller to be working in a propeller stream whose initial velocity, relative to the propeller, is V_2, equal to $(V - kV)$. Then, if $A =$ disc area of propeller, the sectional area of propeller stream before acceleration by the propeller would be $\dfrac{AV_2 + \dfrac{S}{2}}{V_2}$, and the sectional area of the stream after acceleration would be $\dfrac{AV_2 + \dfrac{S}{2}}{V_2 + S}$, so that the stream may be represented diagrammatically by Fig. 6. The velocity of stream in this case *at* the propeller is $V_2 + \dfrac{S}{2}$.

§ 4. "**Negative Slip.**"—Taking a hypothetical case, of a full-bodied ship, where $k = \cdot 15$, $V = 22$ feet per second

GENERAL THEORY

(about 13 knots), and $f = 5$ feet per second, then $V_2 = \cdot 85 \times 22 = 18\cdot 7$ feet per second.

$$V_2 + \frac{f}{2} = 21\cdot 2 \text{ feet per second,}$$

but $V + \frac{f}{2} = PN$, so that the *apparent slip* $= \frac{21\cdot 2 - 22}{21\cdot 2}$

—*i.e.*, $- 3\cdot 8$ per cent.

So that the propeller apparently works with a negative slip. This theory of "apparent negative slip" is due

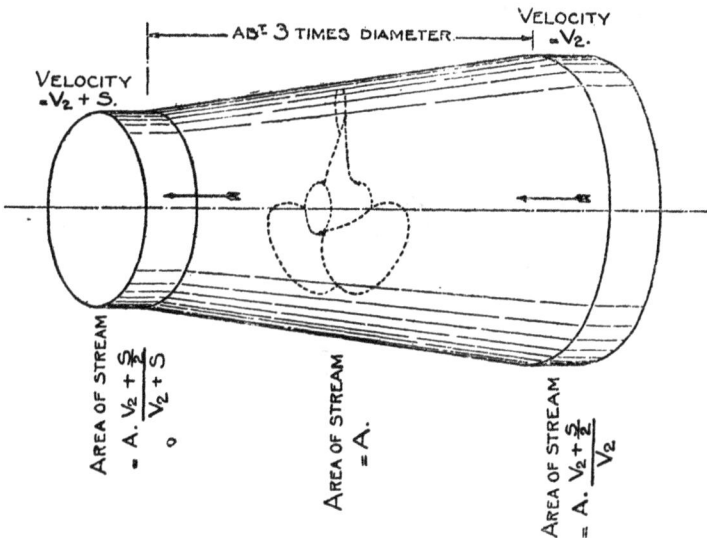

FIG. 6.—DIAGRAMMATIC REPRESENTATION OF PROPELLER STREAM.

to Mr. Barnaby, and seems to be the only satisfactory explanation yet advanced of that phenomenon.

Hitherto, in this chapter, it has been assumed that the propeller stream has a sectional area, at the propeller, equal to the disc area of the propeller; also that all parts of the stream have equal velocity, at any one transaxial plane. Most unfortunately this is not the case. Not only is the sectional area of the stream a very uncertain quantity, but the velocity varies considerably, being greatest at a little distance within the tip-circle of the propeller.

As these indeterminate quantities are of no use to ordinary designers, one is driven to some sort of approximation. The usual assumption made is that the propeller stream has a velocity equal to the product of pitch and revolutions, and an "effective sectional area" bearing a definite relation to the disc area and developed area of the propeller. It appears from a large number of tests

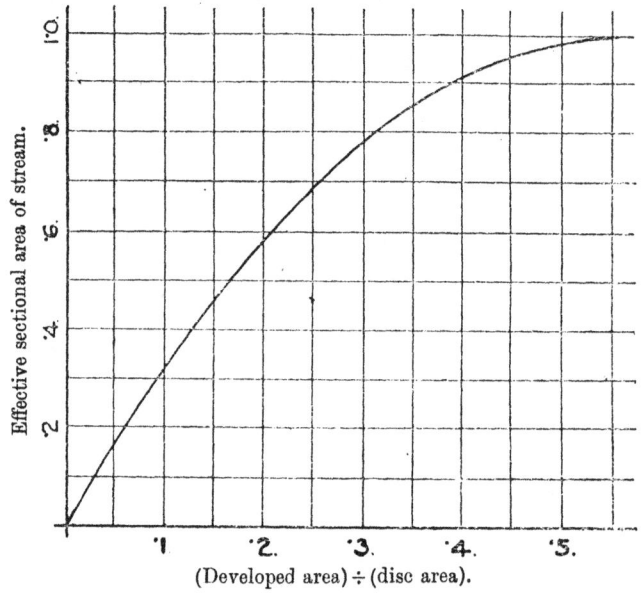

Fig. 7.—Effective sectional area of Propeller Stream.

with models and full-sized propellers carried out by the most prominent investigators, that the definite relation referred to may be represented by the curve, Fig. 7. From this curve it will be seen that a propeller having its developed area equal to about ·55 of its disc area will produce a stream whose effective sectional area is equal to the disc area. Consequently, no increase of thrust can be obtained by increasing the blade area beyond that value.

GENERAL THEORY

It is, however, necessary in certain cases to exceed that value, for reasons discussed under the heading "Cavitation," see § 6.

§ 5. **Derivation of Formulæ.**—It was shown in § 3 that the total work done by the propeller, apart from work done in producing rotation of the propeller stream, overcoming frictional resistance, etc., is represented by $R \times V + \frac{W}{2g}f^2$ foot pounds per second, all symbols having the same meaning as before.

This rate of work may be called "thrust horse-power," or, more correctly, this rate divided by 550. Then we may write

$$550 \times (T.H.P.) = RV + \frac{W}{2g}f^2 = \frac{W}{g}f\left(V + \frac{f}{2}\right).$$

It was also shown that the efficiency of the propeller is measured by $\dfrac{V}{\left(V + \frac{f}{2}\right)}$. Therefore, for a given maximum efficiency f may be expressed as a function of V, and a relation may be stated as follows:

$$T.H.P. \propto WV^2.$$

But $W =$ weight of water delivered per second by propeller; so that

$$W \propto D^2 V.$$

Hence, $T.H.P. \propto D^2 V V^2$, or $D^2 V^3$—

$$i.e., \text{disc area} \propto \frac{T.H.P.}{V^3} \qquad . \quad . \quad . \quad (1)$$

Also V, the speed of the vessel, is proportional to the product PN, so, if we express P as a function of D, then for a given pitch ratio,

$$T.H.P. \propto D^2(NP)^3, \text{ or } D^2 N^3 D^3,$$

$$i.e., D^2 N^3 D^3 \propto D^2 V^3, \text{ from formula 1}.$$

That is to say, $N \propto \dfrac{V}{D}$ . . . . . (2)

42 MARINE SCREW PROPELLERS

Now, writing in place of (1),

$$\text{T.H.P.} = k D^2 V^3,$$

then $k = \dfrac{(\text{T.H.P.})}{D^2 V^3}$. Also, from (2), one may put $D = k_1 \dfrac{V}{N}$, so that $k = \dfrac{(\text{T.H.P.}) N^2}{k_1^2 V^5}$, *i.e.*, $K = \dfrac{(\text{T.H.P.}) N^2}{V^5}$, where $K = \dfrac{k}{k_1^2}$, which is what may be called a "variable constant," or a factor which will have a constant value for cases of closely similar conditions.

Comparative Formulæ.—By the above term is meant formulæ which may be used to determine propeller dimensions by comparison with other previously existing propellers, usually of close similarity to the one under design.

From formula (1) above, we can derive

$$D = K \sqrt{\dfrac{\text{I.H.P.}}{V^3}} \qquad . \quad . \quad . \quad (3)$$

where V = speed of ship; and we may write

I.H.P. $\propto D^2 V$ for constant thrust, or $D^3 N$,

$$i.e., \quad D \propto \sqrt[3]{\dfrac{\text{I.H.P.}}{N}} \qquad . \quad . \quad . \quad (4)$$

$$\text{also } D \propto \sqrt{\dfrac{\text{I.H.P.}}{V}} \qquad . \quad . \quad . \quad (5)$$

and, for the same diameter, developed area $\propto \sqrt{\dfrac{\text{I.H.P.}}{N}}$.

Now, with two vessels of close similarity, if—

V_1 = speed of first vessel;
V_2 = ,, ,, second ,,
I.H.P._1 = power of first vessel;
I.H.P._2 = ,, ,, second ,,
N_1 = revs. per minute, first vessel;
N_2 = ,, ,, ,, second ,,
D_1 = dia. of propeller, first vessel;
D_2 = ,, ,, ,, second ,,

GENERAL THEORY

$$\text{then } D_2 = D_1 \sqrt{\frac{\text{I.H.P.}_2 V_1^3}{\text{I.H.P.}_1 V_2^3}} \quad . \quad . \quad . \quad (6)$$

$$D_2 = D_1 \sqrt[3]{\frac{\text{I.H.P.}_2 N_1}{\text{I.H.P.}_1 N_2}} \quad . \quad . \quad . \quad (7)$$

Basic Formulæ.—By this term is meant formulæ which will serve for finding propeller dimensions without direct comparison with other propellers. It is an unfortunate but indubitable fact that no completely satisfactory formula has yet been evolved to meet the above conditions. Froude's principles, as described and applied by Mr. Barnaby, are entirely logical, and form the basis of practically all accepted systems of propeller design; but these methods permit of an indefinite number of solutions for a given set of conditions, unless the solutions are arbitrarily limited. The reader is referred to Mr. Barnaby's book for a further treatment of the subject.

In order to overcome this difficulty of multiple solutions, charts have been prepared by the writer for the direct determination of propeller diameter and pitch. These charts are based throughout on Froude's principles, but are made to include the case of wide blades, which is nowadays of great importance for turbine-driven vessels.

It may be noted here that in the case of extremely high speed vessels, such as destroyers, fast motor-boats, etc., the propellers are generally considerably smaller than the sizes indicated by accepted theories. Moreover, propellers of larger size, based on such accepted theories, have in sundry well-authenticated instances given distinctly inferior results. A famous shipbuilder has remarked that nine out of ten propellers which are running at the present day would be a great deal better if they were reduced in diameter by 25 per cent. The present writer does not advocate such heroic measures, but he believes this point to be worthy of very careful consideration.

As a matter of interest only, it may be mentioned that

attempts have been made to evolve basic formulæ on the following lines:

Let P = pitch of screw in feet.
A_E = effective area in square feet.
A_D = developed ,, ,, ,, ,,
V = speed of ship in feet per second.
N = revolutions per second of screw.
wV = speed of wake-water relative to propeller in feet per second, where w is a coefficient depending on the form of hull, etc.
S = astern acceleration to be given to propeller stream.
R = tow-rope resistance of ship, at speed V.

Then entering speed of propeller stream = wV; and final speed of stream = wV + S. Assuming, as in § 4, that the speed of stream at the propeller is the mean of these two, and equals the product of (pitch × revolutions), then

$$PN = \frac{2wV + S}{2},$$

$$or \; P = \frac{2wV + S}{2N}.$$

And using the formula $R = \frac{W}{g}f$, f is now written as S, and $W = \frac{Rg}{S}$ pounds per second, where W is the weight of water passed per second by the screw.

Or, in cubic feet per second, $\frac{Rg}{S \times 62\cdot 4}$.

The mean velocity of propeller stream relative to the screw = $wV + \frac{S}{2}$, so that $A_E = \dfrac{Rg}{*62\cdot 4 S\left(wV + \frac{S}{2}\right)}$

Now, if δ be the ratio of developed area to disc area, it was stated in § 4 that $A_E = A_D \sqrt{\frac{\delta}{\cdot 55}}$, so that the above

* Fresh-water value: salt water weighs approximately 64·2 lb. per cubic foot, average value.

GENERAL THEORY

equation can be written $A_D = \dfrac{Rg}{62 \cdot 4S\left(wV + \dfrac{S}{2}\right)} \left(\dfrac{\cdot 55}{\delta}\right)^{\frac{1}{2}}$.

It is, however, clear that the propeller stream is not bounded by the diameter of the screw, but extends beyond the propeller circle, to an extent depending upon the acceleration S, together with the form of hull and position of propeller.

Another source of error lies in the assumption that the face pitch of any propeller is identical with the true pitch. This is not the case. Referring to Fig. 8 (*a*), AD represents a section through a propeller blade which may be regarded as an element of the propelling surface.

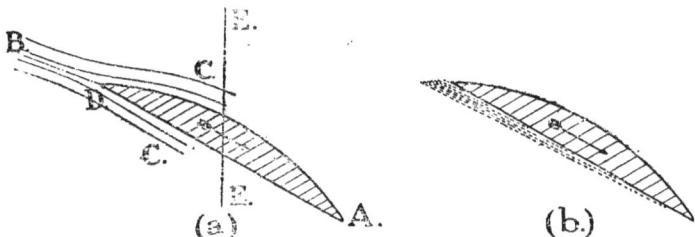

FIG. 8.—EFFECT OF BLADE-THICKNESS AND SURFACE FRICTION.

Imagine that this element is at rest, and that the water flows past it in the direction indicated by the stream lines CC. Then it is quite clear, without any elaborate investigation, that the final path of the water will be in a direction DB, more steeply inclined to the athwartship axis EE than the driving face of the screw. That is to say, that the true pitch is greater than the face pitch. Apart from this, it seems probable that a certain amount of water is carried round with the blades, in the manner suggested in Fig. 8 (*b*), which would also have the effect of increasing the pitch. Sir A. Denny has demonstrated with models in the tank at Dumbarton that a flat sector, perfectly normal to the shaft-line, can have a definite pitch, probably for the above reason.

Again, the actual position of the screw relative to the accelerating position of the propeller stream varies according to the pitch ratio. Coarse-pitched screws appear to have greater "suction" effect than those of finer pitch ratio: such coarse-pitched propellers will therefore take up a position further aft, relative to that accelerating portion of the stream.

From these considerations, it would appear that the formula $A_D = \dfrac{Rg}{62\cdot 4S\left(wV + \dfrac{S}{2}\right)}\left(\dfrac{\cdot 55}{\delta}\right)^{\frac{1}{2}}$ requires so many corrections and adjustments as to be useless for practical design.

§ 6. **Cavitation.**—"Cavitation" is the name given to the formation of spaces filled with air or water vapour at the back of a screw propeller.

In § 3 it was stated that a half or more of the total sternward acceleration of the propeller stream was produced forward of the screw, by means of the suction exerted by the screw. Assuming that one-half of the sternward acceleration is produced forward of the screw, then, if the total thrust of the screw equals the resistance of the ship, equals "R" pounds; then R is made up of $\dfrac{R}{2}$ pounds thrust on the driving face of the screw, and $\dfrac{R}{2}$ pounds suction on the forward face. In order that no partial vacuum may be formed, it is necessary that the water be able to follow the blade with the velocity due to its own head. This velocity $= \sqrt{2gh}$, where h is the depth of immersion of any given portion of the propeller in feet. A large slow-running propeller could be worked efficiently with half of its disc area out of water, provided the above velocity was not exceeded. Where the propeller is completely immersed, as is the case in the great majority of ships, for normal working conditions the allowable velocity becomes $\sqrt{2g(h + 34)}$,

GENERAL THEORY

h being the immersion of the uppermost portion of the screw, and 34 being the head due to atmospheric pressure.

From this it will appear that the suction per unit projected area of the propeller $\left(=\dfrac{R}{2A_p}\right)$ may be as much as 15 pounds per square inch; and the total thrust per unit projected area $\left(=\dfrac{R}{A_p}\right)$ may be up to about 30 pounds per square inch. Unfortunately, the intensity of thrust on the propeller disc is not uniform, being greatest near the circumference (which is the critical point, as shown above). Also, sea-water contains a large proportion of air, dissolved and in suspension; this air is released at low pressures.

Mr. Barnaby states that the maximum permissible intensity of thrust, to avoid cavitation, is $\left(10\cdot 85 + \dfrac{3}{8}h\right)$ pounds per square inch. For this reason it may be necessary to increase δ beyond the value of ·55; although such increase gives no increase of thrust, and actually causes a falling off of efficiency.

In practice, thrusts of 13 to 14 pounds per square inch are common with high-speed vessels, though it is possible that such screws are on the point of " breaking down," or cavitating. It is obvious that the allowable intensity of pressure must be less if the turning moment exerted by the propelling machinery is subject to fluctuation. For this reason, a higher pressure may be allowed with turbine-driven screws (in which the torque is constant) than with propellers driven by reciprocating engines in which the torque may vary in a marked degree. The figure given by Mr. Barnaby may be applied to screws driven by carefully balanced reciprocating engines, while the higher figure given above applies to turbine-driven screws. For 4-stroke cycle internal combustion engines, having four cylinders, the limit is considered to be about 8 to 9 pounds per square inch.

§ 7. **Tip-Speed and Clearance.**—These two quantities are closely related to one another, and to the question of cavitation. Propellers have been run with a tip-speed of 270 feet per second, with a clearance of 16 inches, with no observable ill-effect on hull or blades. In vessels driven by "raised propellers," working in tunnels, where the form of the tunnel follows the line of the propeller circle, the tip-speed has been as much as 105 feet per second, with 3 inches clearance. Again no ill-effect was observed on hull or blades. In these latter cases, however, the screws were not working near the cavitating

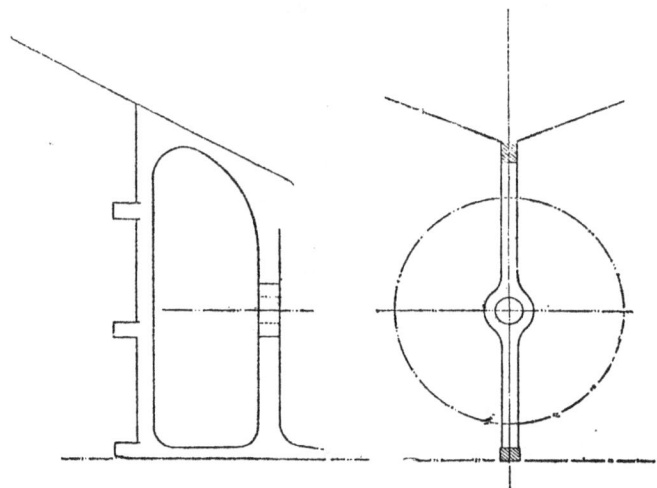

Fig. 9.—Single Screw in Ordinary Stern Frame.

limit. It would appear that a propeller working at or near that limit would cause serious vibration, and possibly damage, if arranged with the line of hull following the screw circle over any appreciable distance. Owing to the condition mentioned towards the end of § 5 a screw running at high tip-speed and mounted in the ordinary type of stern frame (see Fig. 9) should have good tip-clearance.

Overlapping of propeller circles should be avoided.

GENERAL THEORY

Referring to Fig. 10, numbers 1, 2, and 3 screws are respectively right-hand, right-hand, and left-hand. Then the portion "A" of No. 2 screw will be working with more than its normal pitch, owing to the rotation of the propeller-stream from No. 1; while the portion "B" will work with less than its normal pitch, due to the rotation of stream from No. 3. Such a condition is certain to reduce the efficiency of No. 2 screw, and is extremely likely to cause serious vibration and even damage to the centre propeller.

§ 8. **Increasing Pitch.**—The acceleration of the propeller stream was discussed in § 3 and illustrated dia-

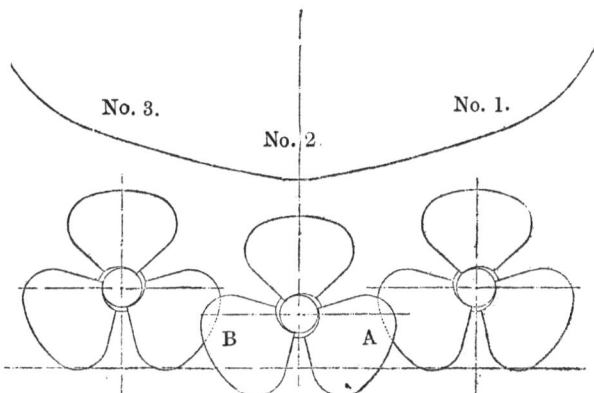

FIG. 10.—EFFECT OF OVERLAPPING OF PROPELLER CIRCLES.

grammatically by Fig. 6. Ordinary mechanical principles of design point to the necessity of varying the pitch of the propeller in order that such part of the acceleration as takes place in the actual region of the blades may be effected without shock or eddying. In other words, the pitch at the leading edge should be less than that at the trailing edge, particularly in blades of great breadth compared with the propeller diameter.

Assuming as in § 3 that the total axial distance in which acceleration occurs is equal to 3 D, and using "B" to represent the projected breadth of one blade at a radius equal to $\frac{\cdot 65 D}{2}$, a formula may be derived as follows:

Axial length of blade at radius noted above $= \dfrac{PB}{\cdot 65\pi D}$

Acceleration of stream relative to entering speed $= \dfrac{S}{wV}$

Increase of pitch in breadth of blade $= \dfrac{\dfrac{S}{wV}P^2B}{6D^2}$

where P is the pitch at the leading edge of the blade.

This formula, however, is based so much on assumption that a variation of pitch calculated by its means may be positively disadvantageous. For this reason it is considered safer to design propellers with a uniform pitch at the driving face, unless it is possible to carry out careful model experiments to verify the design. Turbine-driven propellers for high-speed vessels, though usually of broad-bladed type, are almost invariably of uniform pitch.

The actual form of blade section (with flat driving face and convex back), as illustrated in Plate IV. (*g*), gives the effect of increasing pitch. Reference is made to this point in § 5.

§ 9. Augment of Resistance.—If a ship, without propeller, be towed at a given speed, and then driven by its own propeller at that same speed, the actual driving thrust being measured in each case, it is found in practically all cases that the second thrust is greater than the first. The reason for this is that the backward acceleration of the propeller-stream has an effect on a part of the hull corresponding to an increase of speed of the ship. This means that the effective horse-power necessary to propel a vessel at a given speed must be increased

GENERAL THEORY

beyond the tow-rope horse-power for that speed, by a factor called the "augment of resistance" factor. This correction is sometimes described as "thrust deduction correction." Inasmuch as it has really nothing to do with deduction of thrust, the term seems misleading. If we write $\frac{\text{propeller thrust}}{\text{tow-rope thrust}} = a$, then a is the augment of resistance factor.

§ 10. **Efficiency.**—Suppose a screw propeller to be merely a means of setting water in motion in the manner of a pump. Then,

useful work = work done in accelerating water in a longitudinal direction;

waste work = work done in rotating water, in overcoming frictional resistance; in wave-making, etc.

The ratio of $\frac{\text{useful work}}{\text{total work}}$ may be called "pump efficiency."

Suppose, again, that a quantity of water delivered by a screw is used to propel a ship. Then from § 5, "water efficiency" = $\frac{V}{V + \frac{S}{2}}$, or more fully $\frac{wV}{wV + \frac{S}{2}}$.

It has also been shown that the entering velocity of the propeller stream is affected by the hull, while the hull resistance is in turn affected by the acceleration of the propeller stream. The product $\frac{1}{w} \times \frac{1}{a} = $ "*hull efficiency.*"

The *total efficiency* of the propeller may then be expressed as $\frac{\text{tow-rope horse-power}}{\text{shaft horse-power delivered to screw}} = $ (pump efficiency) × (water efficiency) × (hull efficiency).

Hull Efficiency.—The value of $\frac{1}{w} \times \frac{1}{a}$ is fairly constant, varying from ·95 to 1·0. For twin-screw ships of reasonably fine lines it may be taken as ·97; but it will be lower for vessels with full after-body.

Water Efficiency.—This depends on the value chosen for "S," to suit the speed of revolution of the main engines, and other conditions. The value of "w" has been given by Mr. Doig as follows:

$$w = K - \frac{\text{block coefficient of ship,}}{3}$$

where K = ·99 for single screw,
        1·03 ,, twin screws with bossing,
        1·07 ,, ,, ,, without ,,
        1·01 ,, triple or quadruple screws (average value).

Fig. 11 shows curves derived from the above formula compared with points plotted from actually measured values of w as given by Mr. Barnaby. It will be seen that these actual values hardly warrant any general conclusion. The formula may be used for guidance, but should be checked whenever possible by trial results.

Pump Efficiency.—The exact evaluation of "pump efficiency" requires rather an elaborate mathematical process, which, moreover, must be repeated for all changes of pitch ratio or blade-form. It is also obvious that this pump efficiency depends upon the value of $\frac{S}{V}$. By applying Froude's theory, that any element of a propeller blade can be reckoned as an inclined plane, the efficiency may be obtained for a given set of conditions. Needless to say, the extension of such calculations to a complete propeller is a laborious business. In these circumstances, recourse is had to the results of numerous trials with models and with full-sized screws. Such experimental results are in general agreement with the figures derived from calculation. Fig. 12 shows values of (pump efficiency) × (water efficiency) plotted from results published by a number of investigators.

GENERAL THEORY

§ 11. **Strength of Blades.**—The stresses obtaining in a propeller blade are due to—

(1) Bending moment due to fore-and-aft thrust.

(2) ,, ,, ,, rotation.

(3) Centrifugal force, and—in the case of screws with "set-back" blades—bending moment due to the same.

Now, in settling the thickness of a blade to meet given

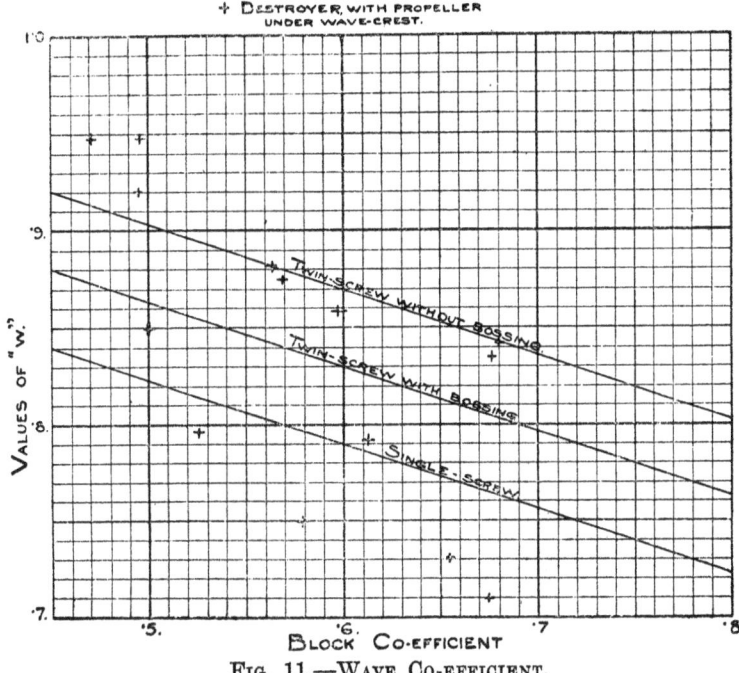

Fig. 11.—Wave Co-efficient.

conditions, we may assume a figure, and then calculate the stresses due to above causes. This method is, of course, one of simple trial and error.

A number of formulæ are also in common use, by means of which the thickness of blade may be calculated. Some of these formulæ are of fairly logical construction, while others are frankly empirical. In any case sizes obtained by the use of such formulæ should as a rule

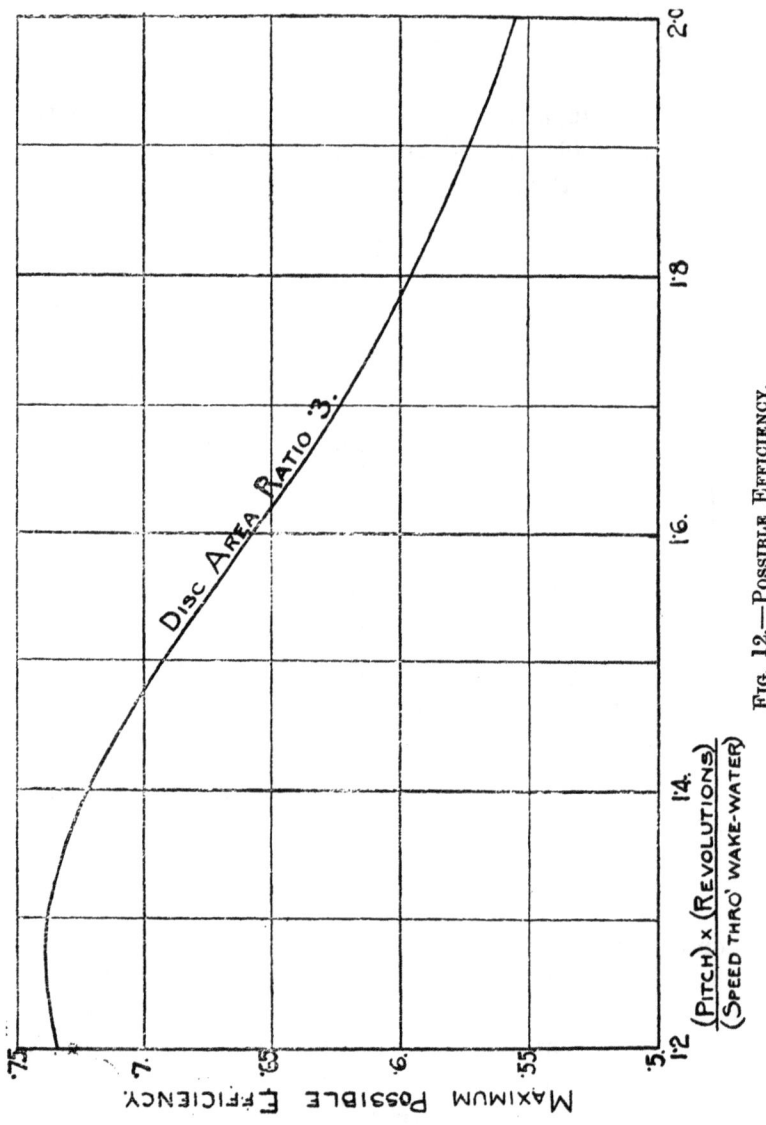

Fig. 12.—Possible Efficiency.

GENERAL THEORY

be checked by full calculation of stresses. A few of the more logical formulæ may be briefly discussed here.

(1) $T_1 = \sqrt{\dfrac{d^3}{n \times b} \times K} + C.$

(2) $T^2 = \dfrac{N \times I.H.P. \times (D - d_1)}{B_1} \cdot \left(\dfrac{d_1}{P \times 3} + \dfrac{20}{R \times D}\right).$

(3) $M = \dfrac{B \times T^2 \times R \times P}{I.H.P. \times (D - d_1)}.$

(4) $T = J\sqrt{\dfrac{I.H.P. \times D}{S \times B}}.$

The meanings of these symbols are as follows:

T = thickness of blade at surface of boss in inches.
T_1 = thickness of blade at centre of shaft in inches; see Plate IV. (*a*).
d = diameter of tail-end shaft in inches.
d_1 = diameter of boss in feet.
D = diameter of propeller in feet.
n = number of blades in one propeller.
b = breadth of blade at root in inches, measured parallel to shaft; see Plate IV. (*b*).
B = breadth of blade at root in inches, measured as in Plate IV. (*c*).
B_1 = breadth of blade at root in inches, measured as in Plate IV. (*d*).
R = revolutions per minute.
P = pitch of propeller in feet.
I.H.P. = indicated horse-power transmitted *through one blade.*
S = speed of vessel in knots.
C, K, J, N, and M are constants with different values to suit various conditions.
T_2 = fore-and-aft thrust on one blade in pounds.
T_3 = athwartship   „   „   „
w = wake correction.

No. 1 is due to Mr. Seaton. It can be expressed in the following form: $T_1^2 b \propto \dfrac{d^3}{n}$. The constant C is, of course, merely a "safety allowance." Now the expression $T_1^2 b$ gives a measure of the moment of resistance of the blade to bending at the root, while $\dfrac{d^3}{n}$ is clearly proportional to the torque per blade. No account, then, is taken of variation in diameter and pitch due to exceptional conditions, while the effects of centrifugal force and of "set-back" are also neglected. Thus this formula, which gives excellent and most reliable results for what may be termed "normal" propellers, is hardly suited for cases presenting unusual features.

No. 2 is due to A. E. Wild. Its derivation is given by Barnaby as follows:

Assume that thrust efficiency $= \cdot 58$, then fore-and-aft thrust per blade

$$= \frac{\cdot 58 \times \text{I.H.P.} \times 33{,}000}{\text{S} \times 101 \cdot 3} = 189 \frac{\text{I.H.P.}}{\text{S}}.$$

Assume also that S.H.P. $= \cdot 82 \times$ I.H.P., then athwartship moment of each blade

$$= \frac{\cdot 82 \times \text{I.H.P.} \times 33{,}000}{2\pi \text{R}} = 4{,}300 \times \frac{\text{I.H.P.}}{\text{R}}.$$

And if centre of pressure be taken at two-thirds radius from the centre, the athwartship thrust per blade becomes $12{,}900 \dfrac{\text{I.H.P.}}{\text{RD}}$.

These two components are combined, and the total bending moment at the root of the blade obtained. It will be seen that this formula, though in a way more complete than No. 1, still neglects the effects of centrifugal force, etc.

No. 3, sometimes called the "Admiralty" formula, is most suitable for comparing screws of fairly similar form. Writing it thus: $BT^2 \propto \dfrac{\text{I.H.P.}}{P \times R}(D - d_1)$, it is seen that the only stress dealt with is that due to bending caused by fore-and-aft thrust.

GENERAL THEORY

No. 4 is derived in a similar manner to No. 3, thus:
Moment of resistance of root of blade

$$= \frac{T^2 B f}{9} = \frac{33{,}000 \times \text{I.H.P.} \times \cdot 5 \times (\cdot 325\, D \times 12 - 6d_1)}{S \times 101 \cdot 3},$$

assuming that thrust efficiency is ·5, and that centre of pressure is ·65 of radius from the centre of the screw. Then if we reckon the average value of d_1 as ·2D,

$$\frac{T^2 \times Bf}{9} = \frac{440\, \text{I.H.P.}\, D}{S},$$

which can be written as—

$$T = J \sqrt{\frac{\text{I.H.P.}\, D}{SB}}.$$

For checking the stresses, after settling upon a thickness in the above manner, the following method is convenient, though perhaps rather long. The symbols have the meanings given above:

$$T_2 = \frac{\cdot 5 \times \text{I.H.P.} \times 33{,}000}{101 \cdot 3 \times w S}.$$

$$T_3 = \frac{\cdot 9 \times \text{I.H.P.} \times 33{,}000}{\cdot 6 \times \pi \times D \times R}.$$

Moment, $M_{T_2} = T_2(4 \cdot 2\, D - 6 d_1).$
$M_{T_3} = T_3(3 \cdot 6\, D - 6 d_1).$

The resultant of these moments is then resolved as shown in Fig. 13, to M_1 and M_2, which are respectively perpendicular and parallel to the working face.

Tensile stress at XY (see Fig. 13) due to M_1

$$= \frac{35}{4} \times \frac{M_1}{BT^2} \quad . \quad . \quad . \quad (1)$$

Compressional stress at Z due to M_1

$$= \frac{105}{8} \times \frac{M_1}{BT^2} \quad . \quad . \quad (2)$$

Tension at X and compression at Y due to M_2

$$= \frac{15 M_2}{B^2 T} \quad . \quad . \quad . \quad . \quad (3)$$

Also, if W = weight of blade outside stress-section, point G (see Fig. 13) is the C.G. of the portion of the blade outside the stress-section; then, the centrifugal force $F = \dfrac{WrR^2}{35,200}$.

F, in turn, may be resolved into F_1 and F_2, as shown in the figure; then tension due to F_1

$$= \frac{F_1}{\text{area of section}}, \text{ or approximately } \frac{F}{2/3 bt} \quad . \quad (4)$$

The moment due to $F_2 = F_2 \times m = Fe = M_F$.

Tension at XY (Fig. 13) due to $M_F = \dfrac{35}{4} \times \dfrac{M_F}{BT^2} \quad . \quad (5)$

Compression at Z due to $M_F \quad = \dfrac{105}{8} \times \dfrac{M_F}{BT^2} \quad . \quad (6)$

Total tensile stress at XY $= (1) + (3) + (4) + (5)$.
,, compressional stress at Z $= (2) + (6) - (4)$.

This system is due to Mr. W. Kerr.

The system preferred and recommended by the writer is fully described in Chapter II. By its use, blade thicknesses are settled and total stresses calculated at the same time. The method is straightforward and need not be discussed here.

§ 12. **Trials and Tank Tests.**—The writer has tried to make clear in this chapter that a number of different propellers may be designed for any given set of conditions, all of which screws might in theory be of about the same efficiency. This being the case, a careful and systematic method of conducting trials becomes of the greatest importance. Although practically every screw-propelled vessel built is put through some form of trial, either before or immediately after being handed over to the owner, yet very few of these trials are put to serious use for gaining knowledge of screw propeller design. Occasionally several designs of propeller are

GENERAL THEORY 59

tested on the same ship, but these tests are seldom carried to a logical conclusion. As a rule, as soon as any noticeable improvement of performance is obtained, such tests are discontinued; this, of course, is almost inevitable under commercial conditions.* These difficulties point strongly to the advantage of tests with model propellers in experimental tanks. The laws of

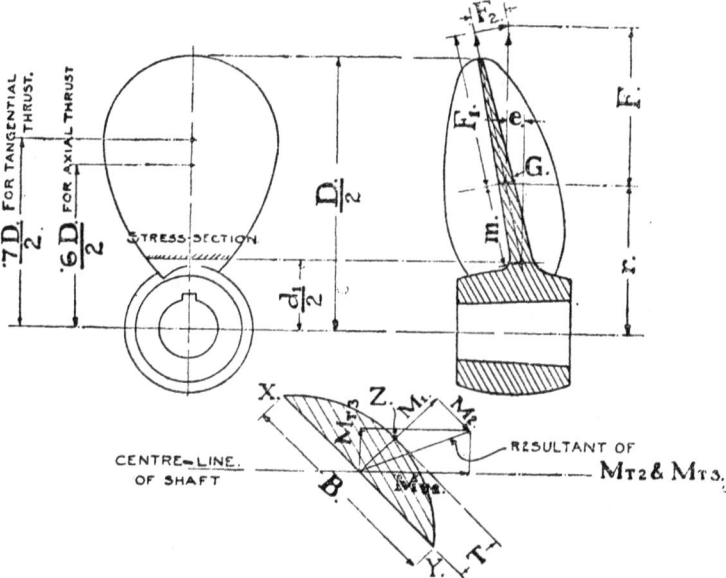

Fig. 13.—Method of Checking Blade Stresses.

comparison for such model screws are well established and may be set forth as follows:

Let the diameters of the two screws be D and d re-

* In the case of vessels of the trawler type, which are built in considerable numbers without variation of design of hull or propelling machinery, systematic tests could very well be carried out, by varying the proportion of screws fitted to successive ships. It may be noted in passing that the important consideration of "using existing patterns" could hardly be urged as an objection to such a series of tests, since propellers of the class under discussion are very seldom cast from patterns.

spectively, and the revolutions per minute be N and n, the speeds V and v, and the thrusts T and t; then if

$$\frac{D}{d} = l \text{ and } V = v\sqrt{l}$$

and the propellers are of similar design and have the same slip,

$$\text{then } n = N\sqrt{l}$$
$$\text{and } T = l^3 \times t.$$

Wherever possible a model screw should be tested in conjunction with the appropriate model hull. The following points may be mentioned as important in the conduct of trials with full-sized screws:

The *accurate* measurement of *power* as well as speed and revolutions, and an accurate measurement of the actual face-pitch of the finished propeller, which is seldom the same as the designed pitch, owing to deformation of castings, etc.

As regards publication of results of tests, it is possible that in the future British engineering firms may take a lesson from their American confrères, and learn the wisdom of making public all information they may possess on subjects of *general* design. Unfortunately, this has been the practice of very few builders in the past.

CHAPTER V

PATTERN-MAKING AND MOULDING

Patterns for normal blades—For set-back blades—For varying pitch—Moulding—Sweeping up for normal blades—For set-back blades—For varying pitch.

IN Chapters I., II., and III. the progress of a propeller has been traced out, from the preliminary stages of design to the final preparation of working drawings. It is now proposed to deal briefly with the processes of pattern-making and moulding. Perhaps it is hardly necessary to say that the casting of propellers is largely a specialized branch of foundry work, and, as such, is in the hands of comparatively few firms. At the same time it is certainly desirable for the designer to take into consideration all the necessary mechanical processes.

§ 1. **Pattern-Making.**—Owing to the heavy cost of pattern-making it is not customary to prepare wooden patterns unless there is a reasonable probability of a considerable number of identical propellers being required. It will be obvious that this condition seldom applies to any but the smallest. Moulds for larger propellers are swept up directly in the sand, in the manner to be described later. The method usually adopted in preparing a pattern for a screw of uniform pitch is as follows:

Staves are prepared of an exact thickness—say 1 inch—and are cut to the shape indicated at (*a*), Fig. 14. On each stave is scribed a radial line AB, the distance BC being carefully set off, equal to $\pi \frac{\text{(diameter)}}{\text{(pitch)}} \times$ (thickness of stave).

Along this line AB are marked off distances AD, DE, etc., giving the positions of the stress-sections shown and dimensioned on the working drawing. The staves are then assembled, as shown at (b), Fig. 14. The edge AC of each stave is made to coincide with the scribed line AB on the stave below it.

The projecting edge AC of each stave is now worked down by hand, until a fair helical surface is produced, on which are exposed the scribed lines AB, and the cross

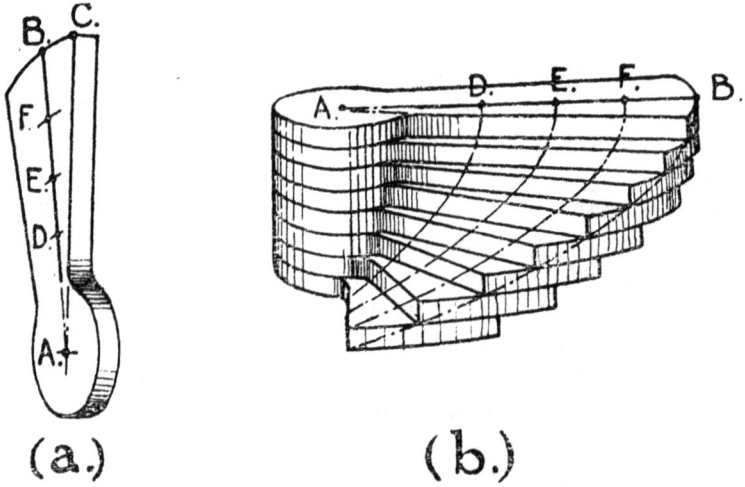

Fig. 14.—Pattern for Normal Blade.

marks D, E, etc. Using these cross marks as guides, the widths of stress-sections are next marked off in accordance with the working drawing. After drawing a fair curve through the points so obtained, the blade is ready for cutting to shape. The back of the blade is then worked down by hand, the thickness being checked by callipers from the working face, to agree with the design.

This process produces a pattern consisting of a boss and one blade. The moulder is thus left to space the blades as required, when preparing the mould. An alternative method occasionally adopted is to prepare the blade pattern, in the manner above described, but separate from the boss. The boss is then adapted to receive the blade pattern in two, three, or four positions according to the number of blades in the finished screw. This, however, entails more work on the patterns, and is of doubtful advantage.

Pattern-making for a set-back blade is not quite as simple as for the normal blade. One method of construction is as follows: Wedge-shaped staves are prepared

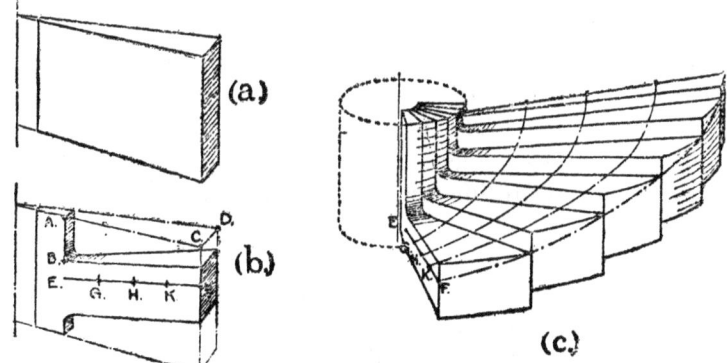

FIG. 15.—CONSTRUCTION OF PATTERN FOR SET-BACK BLADE.

as at (a), Fig. 15, of uniform thickness and taper. The staves are next cut to give the desired set-back of the working face; see Fig. 15 (b). It will readily be understood that the distance AB on any one stave is a multiple of $\frac{\text{CD (pitch)}}{\pi \text{ (diameter)}}$. Parallel to the edge thus prepared of each stave a line is scribed at a distance from the edge equal to $\frac{\text{CD (pitch)}}{\pi \text{ (diameter)}}$. This line is indicated at EF in the sketch.

Cross marks, such as G, H, are also scribed on, to define the positions of stress-sections dimensioned on the working drawing. The staves are then assembled as shown at (c), Fig. 15. The remaining operations are similar to those previously described for the normal blade pattern, save for the additional work of completing the boss, as the staves naturally only form a part of the boss.

Patterns of blades of varying pitch call for careful and intelligent workmanship. The methods described above of building up the pattern in staves can be applied if the working drawings are prepared accordingly. This is probably the most satisfactory solution of the problem. Where this has not been done, but only the form and angle of the various stress-sections given, then probably the best way is to prepare templates, of thin sheet metal or stout strawboard, and work down by hand to fit these templates. The blank for this operation may be prepared in two pieces (see Fig. 16.) This method provides a datum line, which, of course, must be set out at the correct angle, together with positions for the application of the templates.

Pattern-making for the boss of a loose-bladed screw presents no special features, and need not be discussed here. The core box is usually of a simple type, and can nearly always be made on one of the well-known types of core-cutting machines.

§ 2. **Moulding.**—Where patterns are provided, the preparation of moulds for casting propellers presents no special features. The parting of the mould has, of course, to follow the edge of the blade, which necessitates careful ramming of the sand in the bottom half of the mould. This, however, offers no difficulty to a good workman.

If, as is more often the case, the moulds are to be " swept up," then considerable skill and practice are necessary. The appliances necessary are a sweepboard mounted on a suitable pivot, and a pitch-plate or template. This pitch-plate is made of a triangular steel

PATTERN-MAKING AND MOULDING

plate, bent as shown in Fig. 17 (a). The proportions of the triangle, before bending, are as follows:

$$\frac{AC}{BC} = \frac{\text{pitch}}{2\pi \text{ (radius of bend)}}.$$

This pitch-plate being set up concentric with the pivot of the sweepboard (which will be at the centre of the

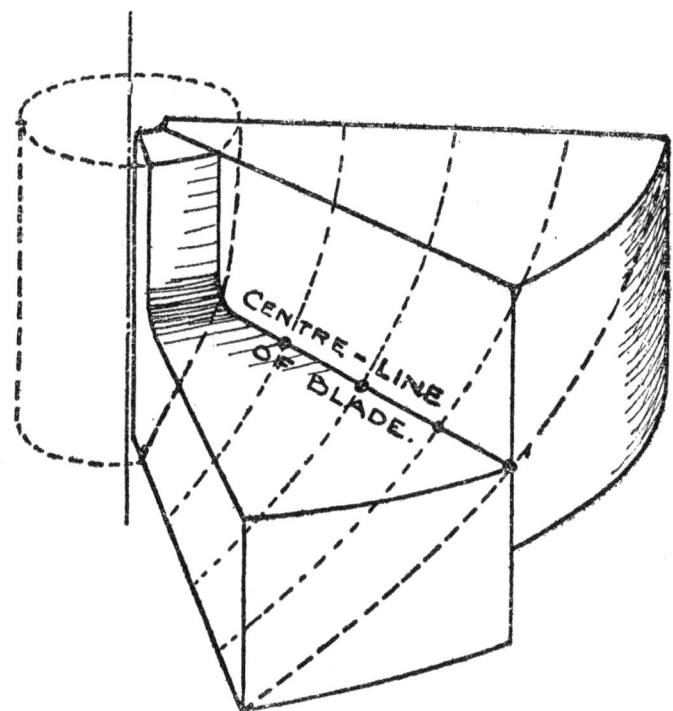

Fig. 16.—Blank prepared for Pattern of Increasing-pitch Blade.

finished mould), a helical surface is swept up in the sand; see Fig. 17 (b). On the sweepboard are marked the positions of stress-sections dimensioned on the working drawing, which may be pricked off on the sand. Laths are prepared of the exact size and shape of the stress-sections (with allowance as necessary, for shrinkage); these laths are pinned down in position. With them as

66 MARINE SCREW PROPELLERS

foundation, the blade is built up in loam (see Fig. 17, c). The outline of the blade is faired by bending a wire or lath round the points fixed by the stress-sections. Parting sand is applied, and the top half of the mould built up in the usual manner. This process, of course, is simply the substitution of a loam pattern for a wood pattern.

Set-back blades may be swept up in a similar manner

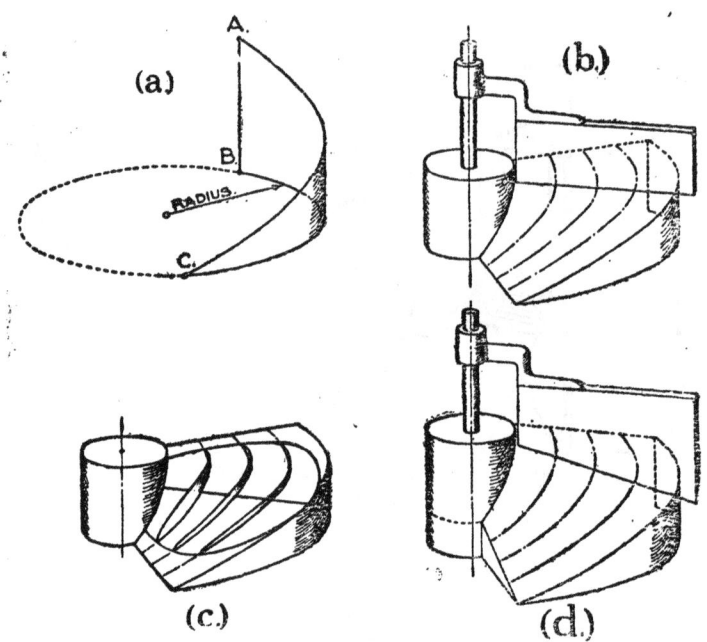

Fig. 17.—Sweeping up a Mould.

by setting the sweepboard to the required angle, as shown in Fig. 17, at (d).

Propellers of large size usually have one or more lifting lugs cast on the blades, which are cut off during the final finishing and balancing process. Care must be taken in arranging these lugs, so that the soundness of the blade casting may not be affected.

In preparing moulds for a propeller of varying pitch,

the first step is to sweep up a true helical surface, and to prick off upon it the positions of stress-sections, as previously described. Two sets of laths are now prepared. The first set gives the contour of the working faces of the stress-sections with reference to a true helical surface, as dimensioned on the working drawing (see Chapter III.). The other set gives the exact shapes and sizes of the stress-sections with shrinkage allowances, as before (see Fig. 18).

The laths of the first set are pinned down in their correct positions, and are used as a foundation for constructing in sand the varying pitch surface desired. Note that this surface is convex in the mould. Again, the

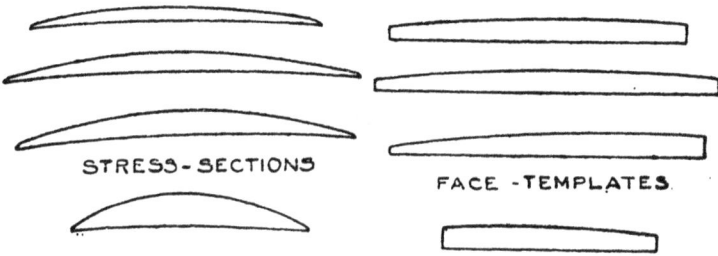

Fig. 18.—Moulder's Templates for Increasing-pitch Propeller.

outline of the blade is faired by means of a wire or a lath.

The laths are left in the mould, to define the positions of the second set of laths, which are now pinned down and used as a foundation for building up the blade in loam. This being done, parting sand is applied, and the top half of the mould built up.

The two halves of the mould having been separated and the loam pattern removed, the laths of the first set (still embedded in the bottom mould) are removed and the surface made good.

On the subject of feeders and risers opinions and practice vary so much that it is almost impossible to give any guidance. One method of arranging them is as shown in Fig. 19, the screw being, of course, cast with its centre line vertical. Single blades of loose-blade propellers may be cast very successfully standing up on end with the flange uppermost. A good-sized riser or header on top of this flange will ensure a sound casting.

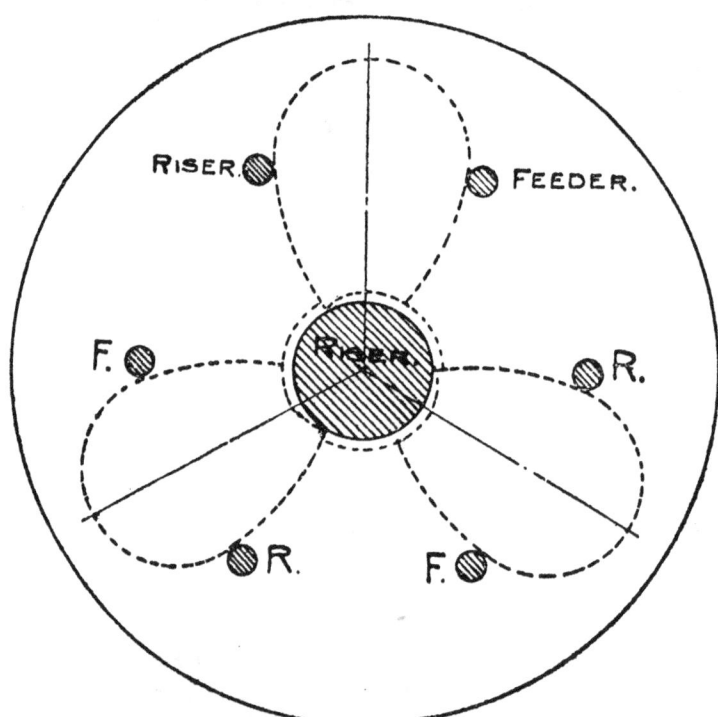

Fig. 19.—Diagrammatic Arrangement of Feeders and Risers.

Owing to the nature of propeller castings and their peculiar liability to deformation in cooling, pouring is an important operation calling for much experience and judgment.

As regards the composition of the metal used, naturally most interest attaches to the subject of bronze for propellers. Most firms undertaking this class of work have a special bronze of their own with a registered name. One reliable mixture has the following composition: copper 88 per cent., tin 10 per cent., manganese 2 per cent.

There is probably little to choose between most of the well-known proprietary bronzes, some of which owe their special features to heat treatment of the finished casting.

CHAPTER VI

MACHINING AND FINISHING

Marking off—Solid propellers—Loose blades—Checking shape—
—Measuring the pitch—Machining the boss—Finishing blade
surfaces—Balancing.

§ 1. Marking Off.—When marking off a propeller casting, due account must be taken of the fact that distortion of the blades in cooling is very common and almost unavoidable. This being the case, it is necessary to mark off the boss (or the flanges of loose blades) so that all blades of the propeller have approximately the same pitch, which is done in the following manner:

Pitch-plates are prepared (as described in Chapter V.), one for each stress-section, dimensioned on the working drawing, and usually one for a section about 1 inch inward from the circumference of the propeller circle. The propeller is set up roughly on a surface table, on which are marked off, with chalk or paint, circles defining the positions of the pitch-plates. Note that the driving face of the blades must be downward—*i.e.*, toward the surface table. The pitch-plates are then applied in their proper positions (see Fig. 20), and the amounts of deviation of each blade from its designed slope at various radii are noted. If the blades are found to be substantially different in pitch from one another, the propeller should be slightly tilted, to equalize the pitches as far as possible. When this has been satisfactorily accomplished, the usual lines can be scribed on the boss, fixing the centre line about which the propeller will ultimately revolve.

MACHINING AND FINISHING

A single loose blade is rather simpler to manipulate. The blade is set up on blocks, roughly in position, and is twisted as necessary to bring its working face as far as possible in conformity with the pitch-plates. In this case more than in that previously described care must be taken that the centre line of the working face is set perpendicular to the flange joint for normal blades, or at the specified angle for set-back blades. This point is of the greatest importance in connection with the balancing of the complete screw.

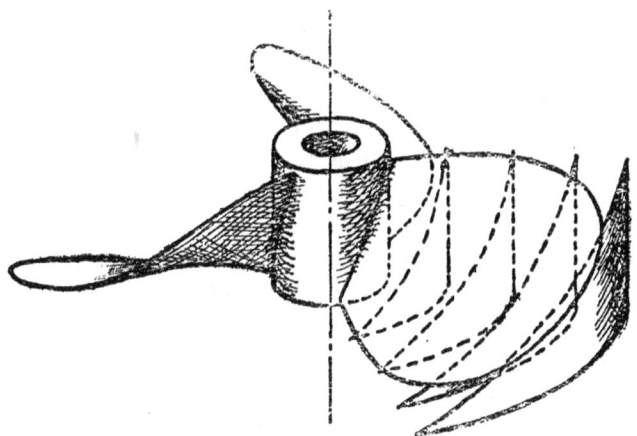

FIG. 20.—PROPELLER WITH PITCH-PLATES IN POSITION.

§ 2. **Checking Shape.**—While the propeller (or single blade, as the case may be) is thus supported on the surface table, the projected shape of the blades may be traced out on the table, or on to a sheet of paper laid on the table, by means of a square or a plumb line. The curve thus obtained serves for checking the shape and area of the blades against the design. This method is not only quicker and easier, but is far more accurate and satisfactory than any attempt to fit paper templates to the blade surface, which at their best can, of course, only be approximate.

Another method of checking the accuracy of the blade form is to scribe on the surface of the blade, from the pitch-plates, the positions of the various stress-sections. The lengths of these sections can then be checked by means of a steel tape. As an alternative, if the pitch-plates are of suitable size the actual length of any stress-section of the blade can be marked off on the appropriate pitch-plate, which can then be measured perhaps more readily than the face of the blade itself.

Except in cases where all the blades have been cast from the same wooden pattern, it is advisable to check the shape of each blade, by one or other of the foregoing methods, to assist in obtaining a proper balance.

§ 3. **Measuring the Pitch.**—Numerous devices are in use for measuring the pitch of a propeller blade at any point on its surface, without the use of pitch-plates. These appliances are commonly known as "pitchometers." A simple form involving no great expense in its construction is illustrated in Fig. 21. It will be seen that the essential parts are (a) a radial arm capable of rotation about the centre line of the propeller boss, and (b) a short straight-edge mounted on a suitable pivot, capable of adjustment to the slope of the blade face at any selected point. In the drawing, instead of a continuous straight-edge, a bar A with two gauge points is shown, hinged at one end to the saddle B. This saddle can be locked in any desired position on the machined angle-bar C, which serves to keep the saddle always in a plane perpendicular to the centre line of the propeller. A steel rule D, laid against the end face E of the saddle, gives a measure of the angle of the bar A, and of the blade face. Now, referring to the line diagram of the apparatus given in Fig. 21, it will be seen that the ratio $\dfrac{\text{KH}}{\text{HG}} = \dfrac{\text{pitch}}{2\pi \text{FG}}$; so that the pitch of the screw at the point measured is equal to $\dfrac{2\pi \text{FG} \cdot \text{KH}}{\text{HG}}$. This at once

MACHINING AND FINISHING 73

suggests that if the length HG be made equal to 6·283 inches (*i.e.*, 2π inches), the pitch of the propeller at the

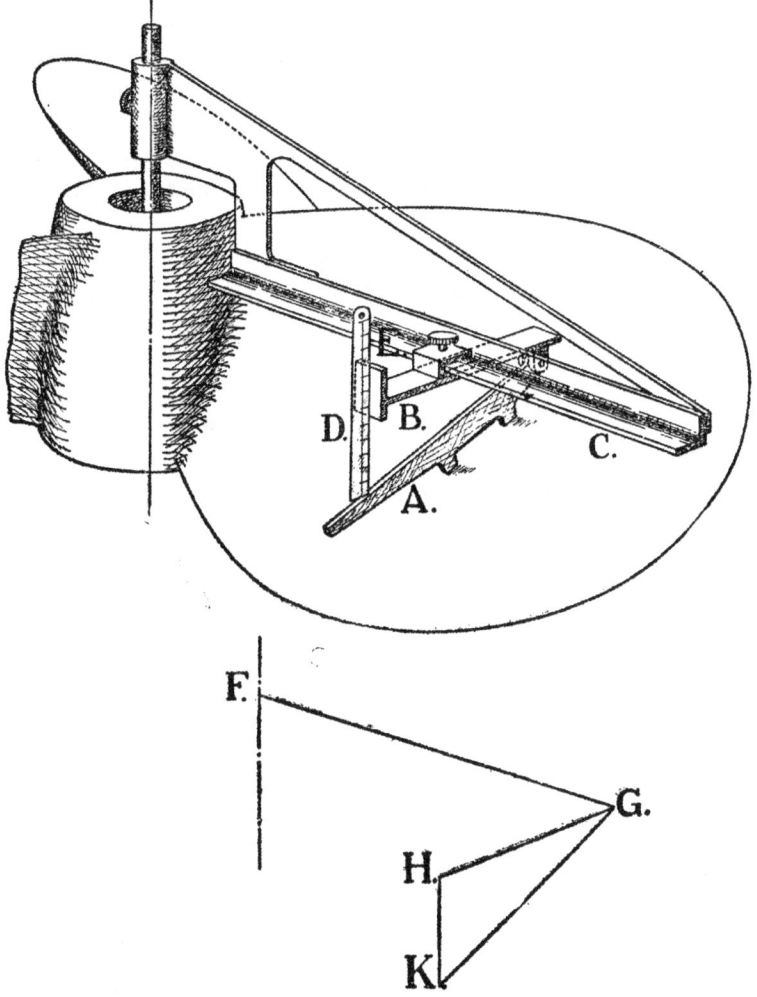

Fig. 21.—Pitchometer.

point tested in inches equals the product of the reading of the rule D, and the distance of the saddle B from the centre of the propeller, both in inches.

§ 4. Machining the Boss.—The type of machine used for the operation of boring out the boss depends upon the size and type of the propeller. The boss of a loose-blade propeller is readily machined on a vertical boring machine, which serves for the operations of facing the boss to receive the blades as well as boring it for the tail-end shaft. Horizontal boring machines of the "Harvey" or similar type are also well adapted for these operations, and are particularly suitable for machining the flanges of the loose blades.

Solid propellers, except those of small size, are usually dealt with more conveniently on the ordinary type of face-plate lathe. Special care must be devoted to the method of bolting the propeller on to the face-plate to avoid strain and distortion of the thinner portions of the blades. Lugs are occasionally cast on the boss, to allow of its attachment to the face-plate without any strain being put upon the blades. These lugs are afterwards removed by chipping, sawing, or slotting.

Gauges for the tapered bore are seldom made of the solid plug type, except where a number of propellers are to be made of identical size, which is usually only the case with small propellers. Templates or plate gauges, as illustrated in Fig. 22, are quite satisfactory for ordinary work, but must be carefully inspected and checked by the tool-maker before and after use.

The slotting of keyways can in most cases be accomplished on an ordinary slotting machine, but is more conveniently performed on a slotting machine of the inverted type, in which the tool-bar is operated from below the table. It may be noted that it is always worth while to spend a little extra time over this operation to eliminate inaccuracies caused by the spring of the cutting tool.

§ 5. Finishing Blade Surfaces.—The machining of propeller blade surfaces is not as yet a common operation. The working face of a uniformly pitched blade may be machined on an ordinary face-plate lathe by fitting suitable special wheels to the change-gear. It will be readily seen, however, that such an operation could not be counted as a practical method. Moreover, the contour of the back of the blade could not be mechanically formed in any satisfactory manner. Special machines have been produced for planing the blade surfaces, in which the contour of the back of the blade is traced out by a series of formers. Such machines are, however, hardly "universal" machines in the sense of being capable of dealing with all types of blade. There

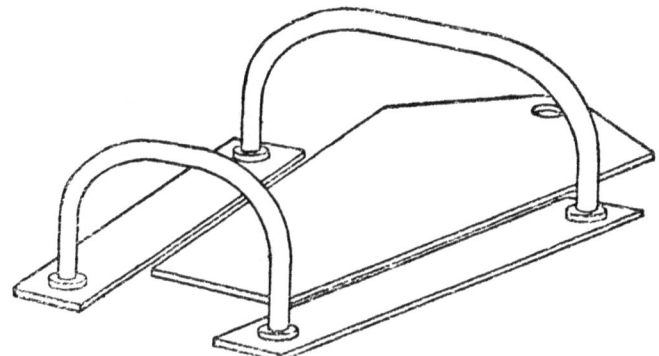

FIG. 22.—TEMPLATES FOR BORE OF PROPELLER BOSS.

is no doubt of the desirability of machining the blade surfaces if it could be done in any reasonable manner, but the problem is certainly not easy of solution, and the market for such an appliance would be too restricted to make it a practical commercial proposition.

In consequence of the difficulties mentioned above the finishing of the blade surfaces is usually carried out by hand. Excess metal (detected by the use of pitch-plates or pitchometer) is removed by chipping where necessary, and the surface of bronze blades may be finished by filing. Cast iron or steel blades are seldom carried to such a high finish, but are merely chipped to balance, and are then painted. Bronze propellers for vessels of moderate speed may be left as received from the foundry—*i.e.*, with sand-blast finish. For speeds of about fifteen knots and over, and for high-revolution speeds, the surfaces are file-finished as described, while for very high speeds the blades may be polished with fine emery

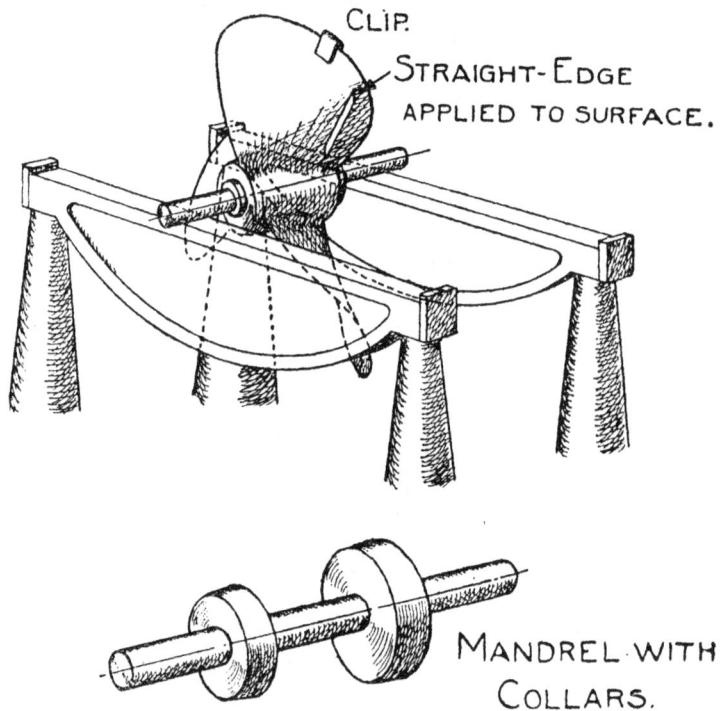

Fig. 23.—Method of Testing Balance.

cloth. It should be remembered that this high polish is not only useful in reducing the frictional losses of the propeller, but also assists in resisting corrosion.

§ 6. **Balancing.**—Careful balancing, which is carried out at the same time as the finishing of blade surfaces, is desirable in all propellers, but is of extreme importance with high-speed propellers. A static balance is all that

MACHINING AND FINISHING

is usually attempted, but in the hands of a skilled workman excellent results may be obtained by this rather primitive process.

The method usually adopted is to prepare a mandrel (see Fig. 23) with loose collars to suit the propeller boss, so that the same mandrel may be used for several different sizes of propeller. Two stout straight-edges are made, usually of cast iron, with very carefully finished edges. These are set exactly level, parallel, and at the same height, and are made of ample strength so as to deflect very slightly under the weight of the propeller. The propeller is then rolled on its mandrel along these straight-edges, when any fault of balance is readily detected. The amount of metal to be removed is found roughly by correcting the balance with clips of known weight, applied to the blades. Metal is then removed as necessary by chipping or filing from any over-thick portion of the blades. Such a thick portion may be located by applying an ordinary straight-edge or rule to the blade surface, as indicated in the drawing, Fig. 23.

CHAPTER VII

REPAIRING

Burning on—Electric welding—Autogenous welding—Replacement—Measurement for replacement—Removing the propeller for repair or replacement.

§ 1. **"Burning On."**—The repairing of screw propellers frequently resolves itself into a question of replacement. The possibilities of actual repair to a damaged blade are distinctly limited. If a blade of bronze, cast iron, or steel be damaged, as shown in Fig. 24, a repair may be effected by "burning on" new metal to replace the portion broken away. This process should not be attempted to any large extent. That is to say, that the metal thus burnt on should only form a small proportion of the complete blade. The blade must be thoroughly heated beforehand, and the section of the new metal should not be much in excess of that of the old metal which it adjoins. The fracture should, if possible, be cut to dovetail form to receive the new metal, and the whole must be very carefully and slowly cooled after burning on.

§ 2. **Electric Welding.**—Damage to the edge of a cast steel blade may be repaired by one of the well-known systems of electric welding. It is inadvisable to attempt this with any but small fractures, as the cost of building up new metal to fill a large fracture is heavy. Careful preparation of the fracture, and highly skilled workmanship on the part of the operator, are absolutely essential. As a rule such repairs should only be attempted to the outermost third of the blade—*i.e.*, the portion lying

REPAIRING 79

above the line AB in Fig. 24. Repairs nearer to the boss may cause serious strains or damage to the main body of metal, rendering it extremely liable to fracture under working conditions.

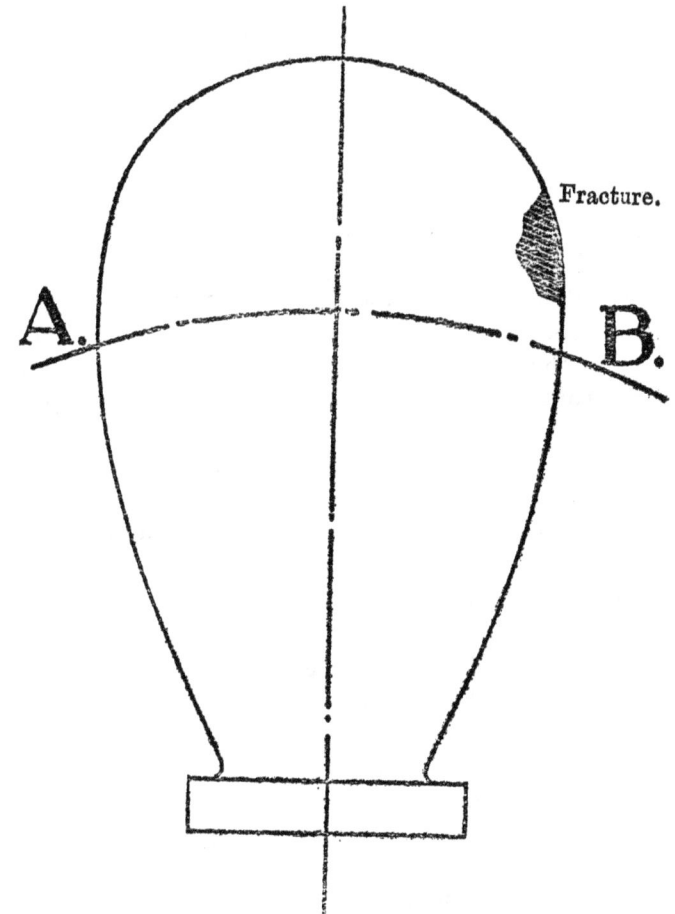

Fig. 24.—Type of Fracture Capable of Repair.

At the risk of repetition, it must again be urged that in electric, as in oxy-acetylene, welding the success of the process is very largely in the hands of the operator;

and only men of proved experience should be trusted with any but the smallest repair of the nature described above.

§ 3. **Autogenous Welding.**—Repairing by means of an autogenous welding process, such as the "Thermit" method, shares with electric welding the considerable merit of being applicable to a propeller in place on its tail-end shaft, while "burning on," of course, entails the removal of the propeller or loose blade and its transport to the foundry, with consequent delay.

Against this advantage must be set the fact that both electric and autogenous welding require special material and appliances, and are only certain of success when carried out by experts in the particular system adopted.

Autogenous welding is perhaps less liable to set up strains in the main body of metal than electric welding, and for that reason can with greater safety be applied to portions of the blades near to the boss.

§ 4. **Replacement.**—Replacement of a propeller or loose blade damaged beyond repair can frequently be effected from the spare gear carried by the ship, while fresh spare gear is ordered from the original builders.

Occasionally, however, it is necessary to have a new blade, or a complete new propeller, cast, for which no drawings are available. In such a case, some system of measurement of the existing propeller is necessary, and it should be possible to make all the measurements required with the propeller in position, for the vessel may have to make a voyage with the damaged propeller while the new one is in course of manufacture.

The length of the boss is determinable by direct measurement, while its diameter, failing suitable callipers, may be found by a light wooden frame as indicated in Fig. 25 (a). By means of such a frame the diameter of the boss is measured at each end and at the centre. If no exact information is available concerning the tail-end shaft its diameter may be assumed as $1\frac{1}{2}$ inches less than

REPAIRING

the diameter of its liner. The taper is then assumed at 1 in 10, and the casting prepared accordingly, with very ample machining allowances. The actual machining

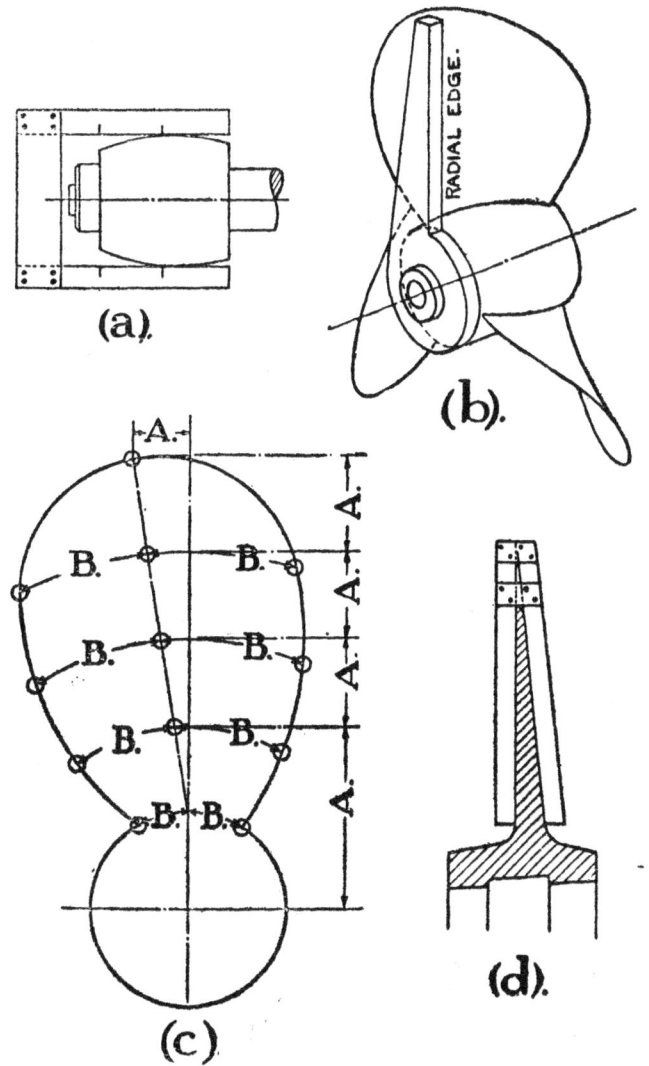

FIG. 25.—MEASUREMENT OF PROPELLER FOR REPAIR OR REPLACEMENT.

of the boss in such a case must be left until the actual sizes of the tail-end shaft can be measured.

For obtaining the particulars of the blades, prepare a board, as in Fig. 25 (b), to rest flat against the after-face of the boss, cut as necessary to clear the tail-end shaft nut.

Now with a rule or tape mark off the centre of width of the blade at several points. Join the points so obtained with a chalk line, which is the centre line of the

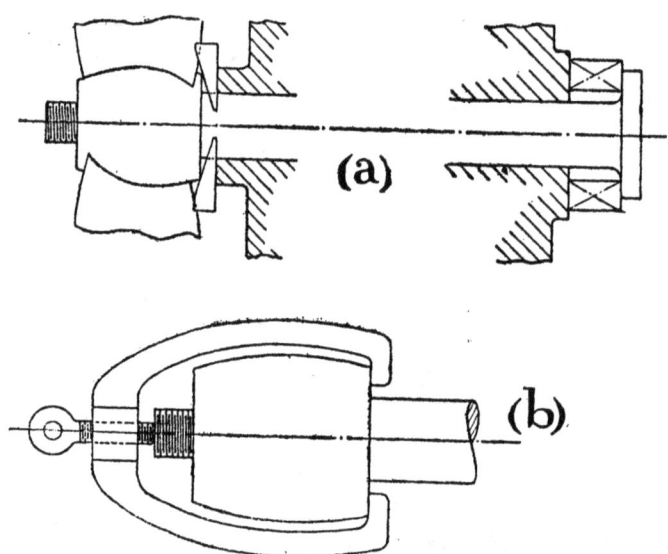

Fig. 26.—Methods of withdrawing Propellers from Shafts.

blade; if the line is not straight (neglecting, of course, small irregularities) join its ends by a straight line, by stretching a chalk line, or by any other convenient means.

With the radial board, a square, and a rule, measure off the position of the blade centre line relative to a radial line. Note that this operation gives the amount of set-back, if any, as well as any "lag" of the blade, such as is shown in exaggerated form in Fig. 3 (b). At the same time mark off on the radial board the extreme

REPAIRING

radius of the blade. Then mark on the board at convenient distances apart positions for stress-sections. With the square, mark these on the blade centre line and on the leading and trailing edges.

A sketch can now be prepared as in Fig. 25 (c). On it are to be noted the distances forward of the radial board, of each of the points denoted by small circles. The dimensions marked A are obtained as above described, and filled in on the sketch. Dimensions marked B, which give the widths of stress-sections, are to be measured on the blade surface. Finally, the blade thickness, if no suitable callipers are available, can be measured by means of two laths as at (d), Fig. 25, fitted respectively to the back and front of the blade.

§ 5. **Removing the Propeller for Repair or Replacement.**—Propellers are commonly removed from their shafts in the manner diagrammatically illustrated in Fig. 26, at (a). The coupling at the forward end of the tail-end shaft is blocked firmly into position, and wedges are inserted between the propeller boss and the after end of the stern tube, and driven home to start the propeller from the shaft. Should the propeller be very difficult to remove, it may be heated by a gas or oil flare, or even by a fire built underneath it. These are primitive methods, and are sometimes avoided by the use of a starting clamp or extractor as illustrated at (b) in Fig. 26. After removing the propeller nut, the clamp is applied, and screwed up tightly. A smart blow or two at the forward end of the boss will then serve to start the propeller from its seat. This device is only suitable for use with comparatively small propellers.

APPENDIX

DEFINITIONS

Apparent Slip: The apparent slip of a propeller, which is the "slip" commonly measured on trials, is the following ratio:

$$\frac{(\text{pitch in feet}) \times (\text{revolutions per minute}) - (\text{speed of vessel in feet per minute})}{(\text{pitch in feet}) \times (\text{revolutions per minute})}.$$

It is usually expressed as a percentage.

Back of Blade: The back of a propeller blade is the foremost surface of the blade.

Developed Area: The developed area of a blade is the true area of the driving face without reference to its shape. The developed area of the propeller is the sum of the developed areas of all its blades.

Diameter: The diameter of the circle swept out by the blade-tips is called the propeller diameter.

Disc Area: The area of the circle swept out by the blade-tips is called the disc area.

Disc-Area Ratio: The disc-area ratio is the ratio of the developed area to the disc area of the propeller.

Driving Face: The driving face of a propeller blade is the aftermost surface of the blade.

Effective Pitch: The effective pitch of a blade is such that, when no thrust is developed, (effective pitch in feet) × (revolutions per minute) = (speed of screw through wake water in feet per minute). The average value of the effective pitch may be taken as 1·02 (face pitch)

Increasing Pitch: A blade of increasing pitch is one in which the pitch increases from the leading edge to the trailing edge. The "mean pitch" of such a blade is the mean of the pitches at the leading and trailing edges.

APPENDIX

Leading Edge: The leading edge of a propeller blade is the foremost edge—that which cuts the water when the propeller is running ahead.

Pitch: The pitch of any point on the driving surface of a blade is the advance per revolution of a true helix co-axial with the propeller and tangential to the blade at that point.

Projected Area: The projected area of a blade is the area of the projection of the blade on a plane perpendicular to the centre line of the propeller shaft. The projected area of the propeller is the sum of the projected areas of all its blades.

Root-Thickness: The root-thickness of a blade is its maximum thickness at the surface of the boss, measured perpendicular to the driving face, and neglecting fillets.

Set-Back: Set-back blades are those in which the centre line of the blade at the tip is abaft the centre line at the root. The axial distance by which the tip is set aft of the root is called the amount of set-back.

Trailing Edge: The trailing edge of a blade is the aftermost edge.

True Slip: The true slip of a propeller is the following ratio:

$$\frac{(\text{effective pitch in feet}) \times (\text{revolutions per minute}) - (\text{speed of screw through wake water in feet per minute})}{(\text{effective pitch in feet}) \times (\text{revolutions per minute})}.$$

Uniform Pitch: A blade of uniform pitch is one of which all portions have the same pitch, as previously defined.

Wake-Coefficient: The wake-coefficient of a vessel is the following ratio:

$$\frac{(\text{speed of stream entering propeller, relative to propeller})}{(\text{speed of vessel through water})}.$$

TABLES

TABLE IV.

Areas of Circles.

Diameter.	0·0	0·2	0·4	0·6	0·8
0	0·00	0·03	0·13	0·28	0·50
1	0·78	1·13	1·54	2·01	2·54
2	3·14	3·80	4·52	5·31	6·16
3	7·07	8·04	9·08	10·18	11·34
4	12·57	13·85	15·21	16·62	18·10
5	19·63	21·24	22·90	24·63	26·42
6	28·27	30·19	32·17	34·21	36·32
7	38·48	40·72	43·01	45·36	47·78
8	50·3	52·8	55·4	58·1	60·8
9	63·6	66·5	69·4	72·4	75·4
10	78·5	81·7	84·9	88·2	91·6
11	95·0	98·5	102·1	105·7	109·4
12	113·1	116·9	120·8	124·7	128·7
13	132·7	136·8	141·0	145·3	149·6
14	153·9	158·4	162·9	167·4	172·0
15	176·7	181·4	186·3	191·1	196·1
16	201·1	206·1	211·2	216·4	221·7
17	227·0	232·4	237·8	243·3	248·8
18	254·5	260·2	265·9	271·7	277·6
19	283·5	289·5	295·6	301·7	307·9
20	314·2	320·5	326·8	333·3	339·8
21	346·4	353·0	360·0	366·4	373·2
22	380·1	387·1	394·1	401·2	408·3
23	415·5	422·7	430·0	437·4	444·9
24	452·4	460·0	467·6	475·3	483·0
25	490·9	498·8	507	515	523
26	531	539	547	556	564
27	573	581	590	598	607
28	616	625	633	642	651
29	661	670	679	688	697
30	707	716	726	735	745

TABLE V.

Squares.

Number.	0·0	0·2	0·4	0·6	0·8
0	0	0·04	0·16	0·36	0·64
1	1	1·44	1·96	2·56	3·24
2	4	4·84	5·76	6·76	7·84
3	9	10·24	11·56	12·96	14·44
4	16	17·64	19·36	21·16	23·04
5	25	27·04	29·16	31·36	33·64
6	36	38·44	40·96	43·56	46·24
7	49	51·8	54·8	57·8	60·8
8	64	67·2	70·6	74·0	77·4
9	81	84·6	88·4	92·2	96·0
10	100	104·0	108·2	112·4	116·6
11	121	125·4	130·0	134·6	139·2
12	144	148·8	153·8	158·8	163·8
13	169	174·2	179·6	185·0	190·4
14	196	201·6	207·4	213·2	219·0
15	225	231·0	237·2	243·4	249·6
16	256	262·4	269·0	275·6	282·2
17	289	295·8	302·8	309·8	316·8
18	324	331·2	338·6	346·0	353·4
19	361	368·6	376·4	384·2	392·0
20	400	408·0	416·2	424·4	432·6
21	441	449·4	458·0	466·6	475·2
22	484	492·8	502	502	520
23	529	538	548	557	566
24	576	586	595	605	615
25	625	635	645	655	666
26	676	686	697	708	718
27	729	740	751	762	773
28	784	795	807	818	829
29	841	853	864	876	888
30	900	912	924	936	949

TABLE VI.

Square Roots.

Number.	Square Root.	Number.	Square Root.
0·20	0·447	0·40	0·632
1	0·458	1	0·640
2	0·469	2	0·648
3	0·480	3	0·656
4	0·490	4	0·663
5	0·500	5	0·671
6	0·510	6	0·678
7	0·520	7	0·686
8	0·529	8	0·693
9	0·539	9	0·700
0 30	0·548	0·50	0·707
1	0·557	1	0·714
2	0·566	2	0·721
3	0·574	3	0·728
4	0·583	4	0·735
5	0·592	5	0·742
6	0·600	6	0·748
7	0·608	7	0·755
8	0·616	8	0·762
9	0·624	9	0·768

TABLE VII.

Fifth Powers.

Number.	Fifth Power.	Number.	Fifth Power.
5	3,125	23	6,436,343
6	7,776	24	7,962,624
7	16,807	25	9,765,625
8	32,768	26	11,881,376
9	59,049	27	14,348,907
10	100,000	28	17,210,368
11	161,051	29	20,511,149
12	248,832	30	24,300,000
13	371,293	31	28,629,151
14	537,824	32	33,554,432
15	759,375	33	39,135,393
16	1,048,576	34	45,435,424
17	1,419,857	35	52,521,875
18	1,888,568	36	60,466,176
19	2,476,099	37	69,343,957
20	3,200,000	38	79,235,168
21	4,084,101	39	90,224,199
22	5,153,632	40	102,400,000

TABLE VIII.
Cube Roots.

Units.	0	1	2	3	4	5	6	7	8	9	Tens.
—	0	1·000	1·260	1·442	1·587	1·710	1·817	1·913	2·000	2·080	0
-teen ..	2·154	2·224	2·289	2·351	2·410	2·466	2·520	2·571	2·621	2·668	1
Twenty-	2·714	2·759	2·802	2·844	2·884	2·924	2·962	3·000	3·036	3·072	2
Thirty-	3·107	3·141	3·175	3·208	3·240	3·271	3·302	3·332	3·362	3·391	3
Forty-	3·420	3·448	3·476	3·503	3·530	3·557	3·583	3·609	3·634	3·659	4
Fifty-	3·684	3·708	3·732	3·756	3·780	3·803	3·825	3·849	3·871	3·893	5
Sixty-	3·915	3·936	3·959	3·979	4·000	4·021	4·041	4·062	4·082	4·102	6
Seventy-	4·121	4·141	4·160	4·179	4·198	4·217	4·236	4·254	4·273	4·291	7
Eighty-	4·309	4·327	4·344	4·362	4·380	4·397	4·414	4·431	4·448	4·465	8
Ninety-	4·481	4·498	4·514	4·531	4·547	4·563	4·579	4·595	4·610	4·626	9
100	4·642	4·657	4·672	4·688	4·703	4·718	4·733	4·747	4·762	4·777	10

TABLE IX.

Knots, Miles, Kilometres

Knots.	Miles per Hour.	Kilometres per Hour.
5	5·7576	9·266
6	6·9091	11·119
7	8·0606	12·927
8	9·2121	14·825
9	10·3636	16·678
10	11·5151	18·532
11	12·6667	20·385
12	13·8182	22·238
13	14·9697	24·091
14	16·1212	25·944
15	17·2727	27·797
16	18·4242	29·651
17	19·5757	31·504
18	20·7273	33·357
19	21·8788	35·210
20	23·0303	37·063
21	24·1818	38·916
22	25·3333	40·769
23	26·4848	42·623
24	27·6364	44·476
25	28·7879	46·329
26	29·9394	48·182
27	31·0909	50·035
28	32·2424	51·889
29	33·3939	53·742
30	34·5454	55·595
31	35·6970	57·448
32	36·8484	59·302
33	38·0000	61·154
34	39·1515	63·008
35	40·3030	64·860
36	41·4546	66·714
37	42·6060	68·568
28	43·7576	70·421
39	44·9091	72·274
40	46·0606	74·128

INDEX

ACCELERATION of propeller stream, 38
Allowable thrust, 47
Apparent slip, 3
— — (definition), 85
Area, blade, for thrust, 8
Areas of circles, table, 86
Augment of resistance, 50
Autogenous welding, 80

Back of blade (definition), 84
Balancing, 76
Basic formulæ, 43
Blade flanges, 19
— form, 22, 31
— — checking, 71
— material, 13
— strength of, 53
— stress on, 17
— surface, 74
— — measurement of, 80
— thickness, 12, 15
Blades, number of, 7
Boss, contour of, 30
— length of, 14
— machining, 74
— recess in, 30
Boss, thickness of, 19
Bronze, 69
Burning on, 78

Casting, 67
Cavitation, 46
Circles, areas, of, table, 86
Clearance, 48
Coefficients, 4
Comparative formulæ, 42
Comparison, design by, 2
— laws of, 60
Cone, contour and construction, 30
Cube roots, table, 89

Definitions, 84
Design by comparison, 2
— complete procedure of, 11
— formulæ for, 2
— from charts, 3
Developed area (definition), 84
Diameter, 3
— (definition), 84
Dimensions, 29
Disc area, 4
— — (definition), 84
Disc - area ratio (definition), 84
Distortion, 70
Drawings, 29
Driving-face (definition), 84
Driving-keys, 14

Effective pitch, 45
— — (definition), 85
Efficiency, 10, 37, 51
Electric welding, 78
Elliptical blade, 23
Extractor, 83

Fifth powers, table, 88
Form of blades, 22, 31
Formulæ, basic, 43
— comparative, 42
— derivation of, 41
— for design, 2
— for scantlings, 12
Frictional losses, 13
Frictionless ship, 33

Gauges for taper of boss, 74

Hull efficiency, 51

Increasing pitch, 49
— — (definition), 85

Keys, 14
Keyways, 74
Knots, miles, kilometres, table, 90

Laws of comparison, 60
Leading edge (definition), 84
Length of boss, 14
Lifting lugs, 66
Limited diameter, 6
Loose blades, 19

Marking-off, 70
Material of blades, 13
Measurement of propellers, 80
Moulding, 64
Multiple propellers, 48

Negative slip, 38
Number of blades, 7

Overlapping of propeller circles, 48

Patterns, increasing pitch, 64
— normal, 61
— set-back, 63
Peripheral speed, 6
Pitch (definition), 84
— effective, 45
— — (definition), 85
— increasing, 49
— measurement of, 72
— plate, 25, 64
Pitchometer, 72
Projected area (definition), 84
Projection, explanation of, 25
— increasing pitch, 28
— normal, 24
— set-back, 26
Propeller stream, acceleration of, 37
— — effective area of, 40
Pump efficiency, 52

Raised propellers, 48
Recess in boss, 30

Removal of propeller, 83
Repair by burning on, 28
— — autogenous welding, 80
— — electric welding, 78
— — replacement, 80
Resistance, augment of, 50
— hull, 34
Root thickness (definition), 84

Scantlings, formulæ for, 12
Set-back, 24
— — (definition), 85
Shaft-horse-power, 9
Slip, 33
— apparent, 3
— negative, 38
Square roots, table, 88
Squares, table, 87
Steel blades, 13
Stress on blades, 17
— — keys, 15
Stress-sections, 17
— — shape of, 29
Studs for loose blades, 19
Sweep-board, 64

Taper of boss, 14
Thermit welding, 80
Thickness of blade, 12, 15
— — boss, 14
Thrust allowable, 8, 47
— deduction correction, 50
Tip speed, 48
Trailing edge (definition), 84
Trials and tank tests, 58
True slip (definition), 85

Uniform pitch (definition), 85

Wake coefficient, 10, 53
— — (definition), 85
Water efficiency, 52
Waveless ship, 35
Welding, autogenous, 80
— electric, 78
Working drawings, 29

PLATE IV.

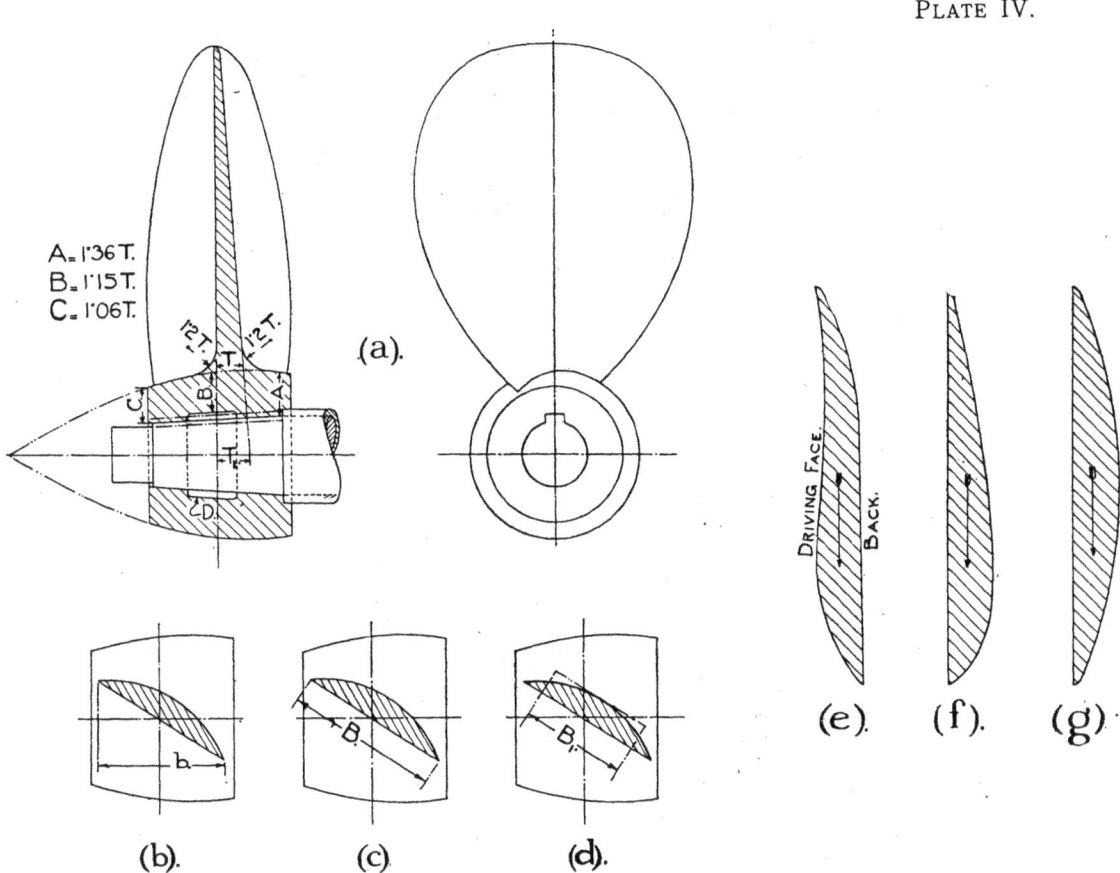

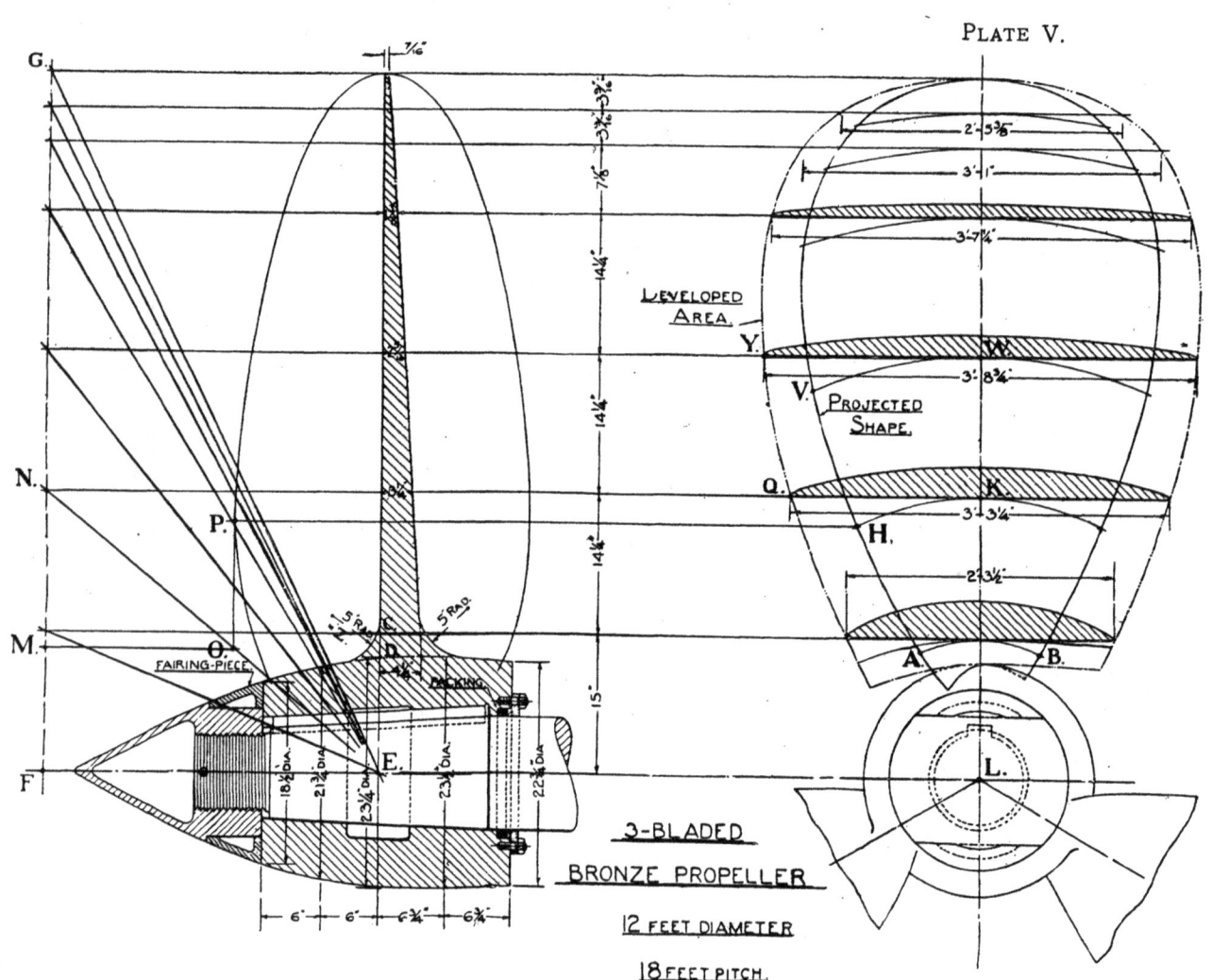

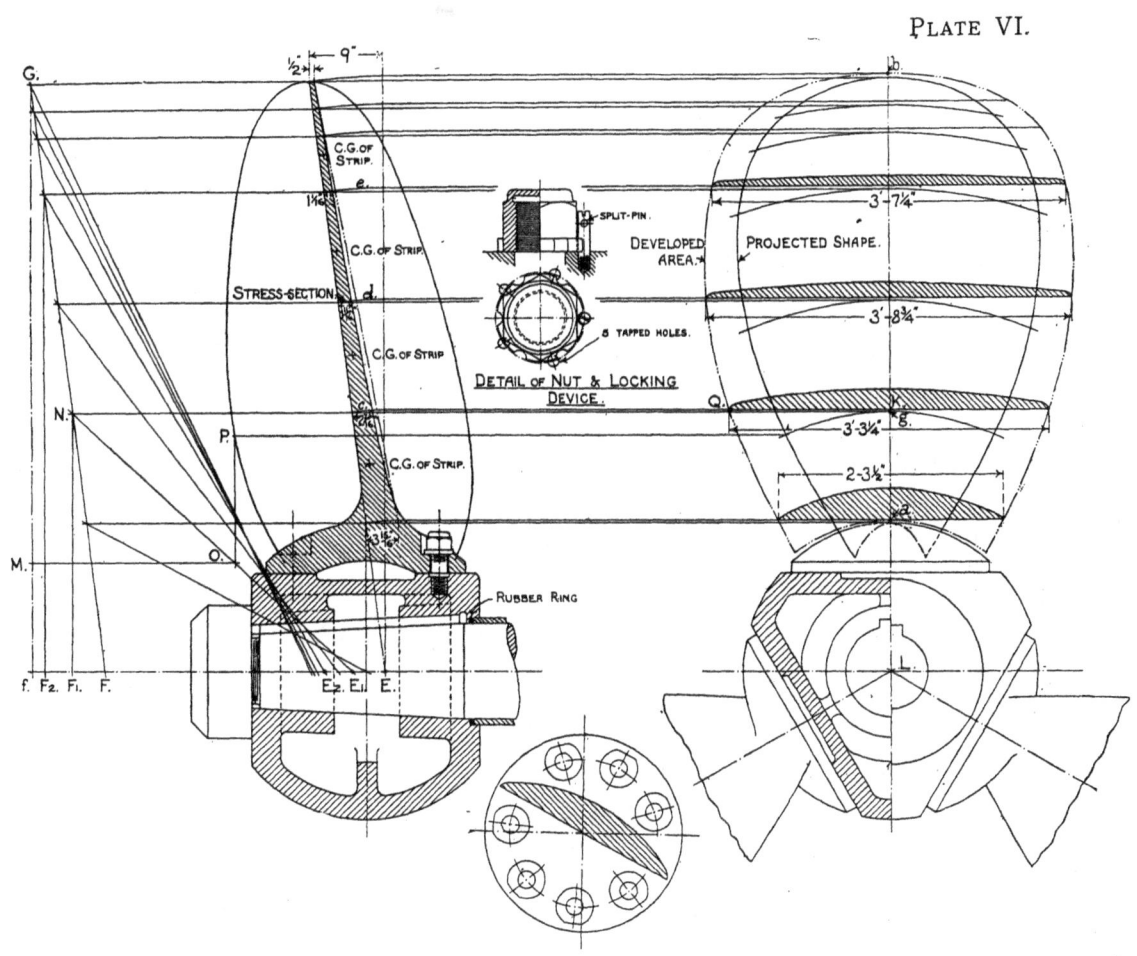

PLATE VI.

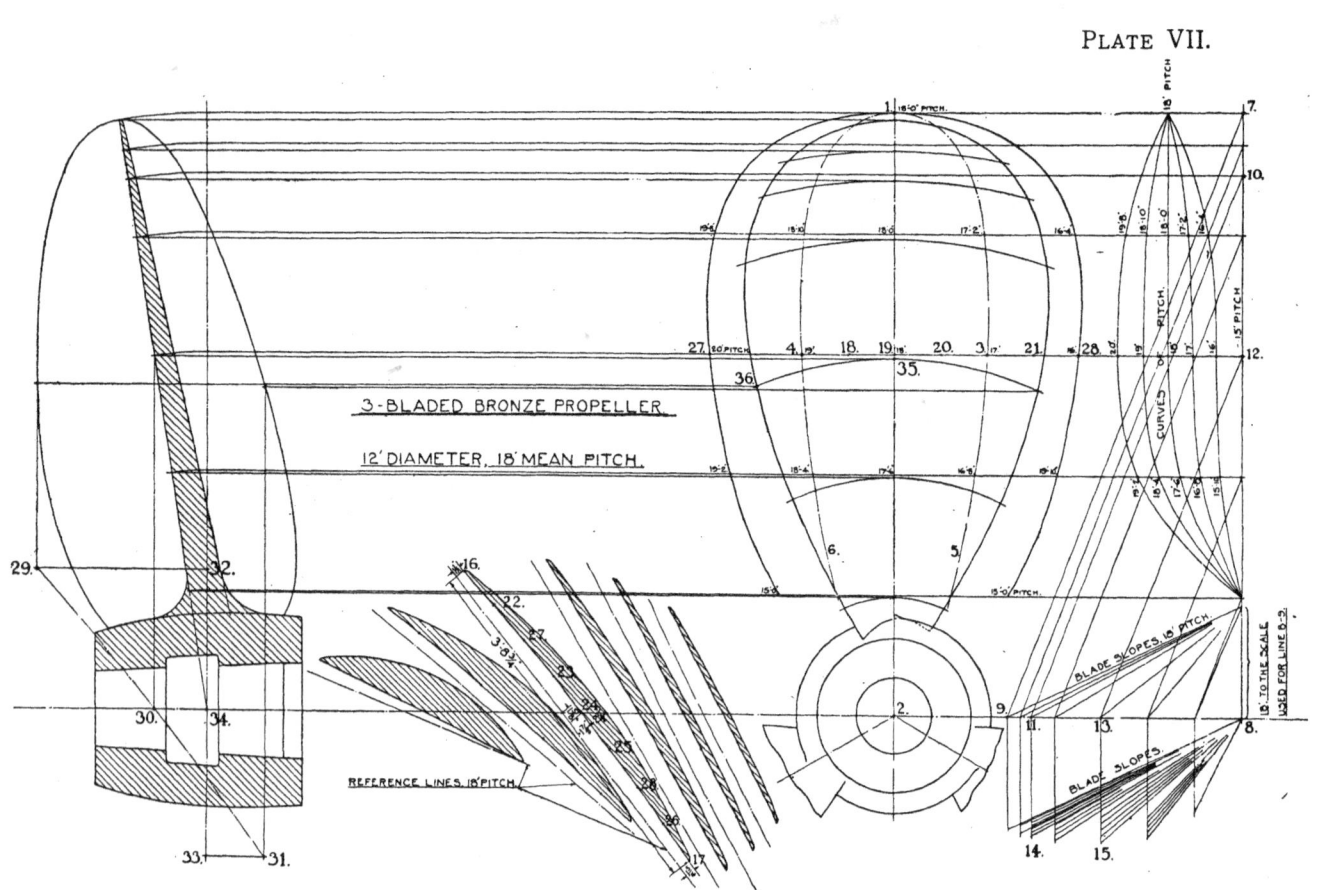

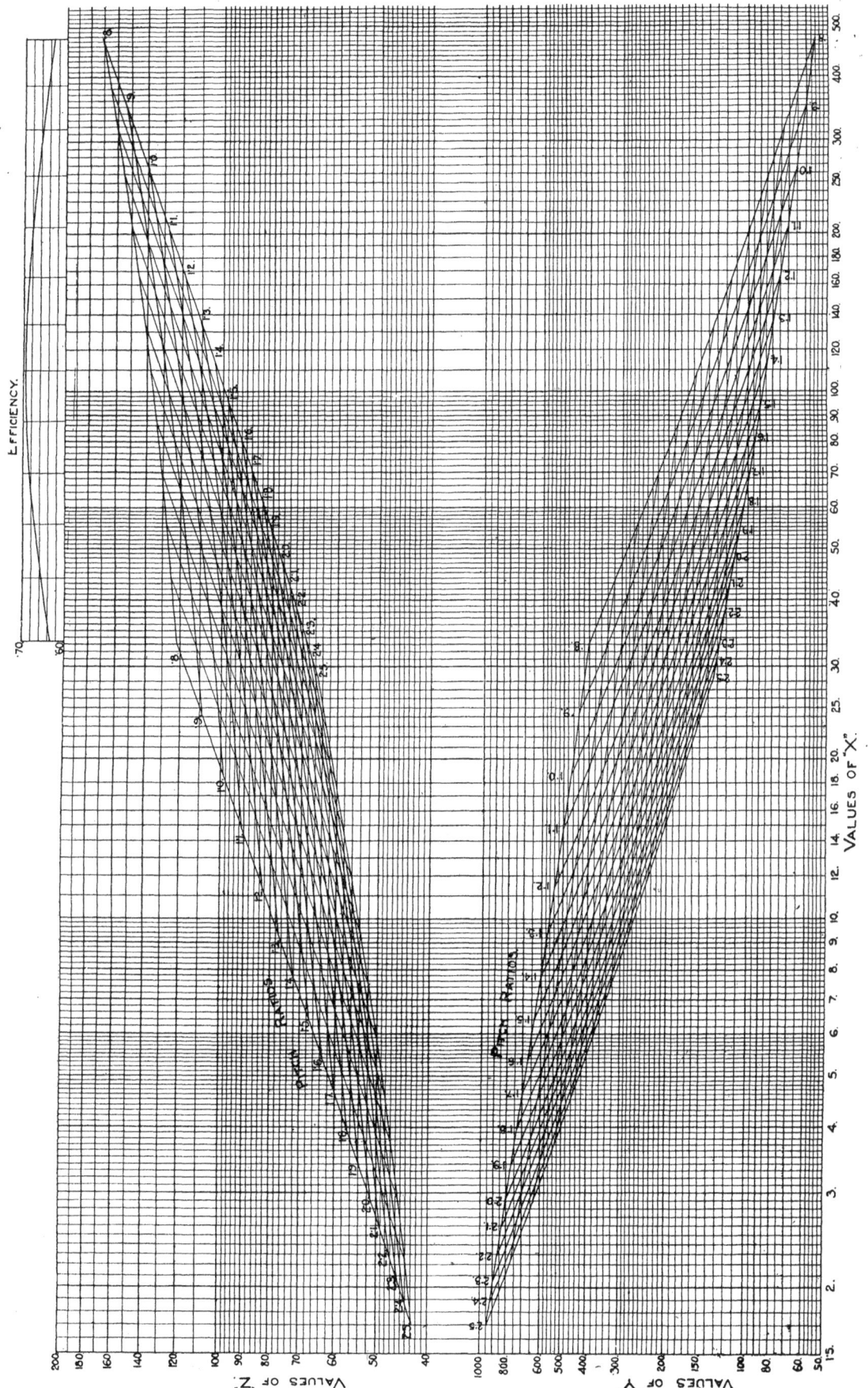

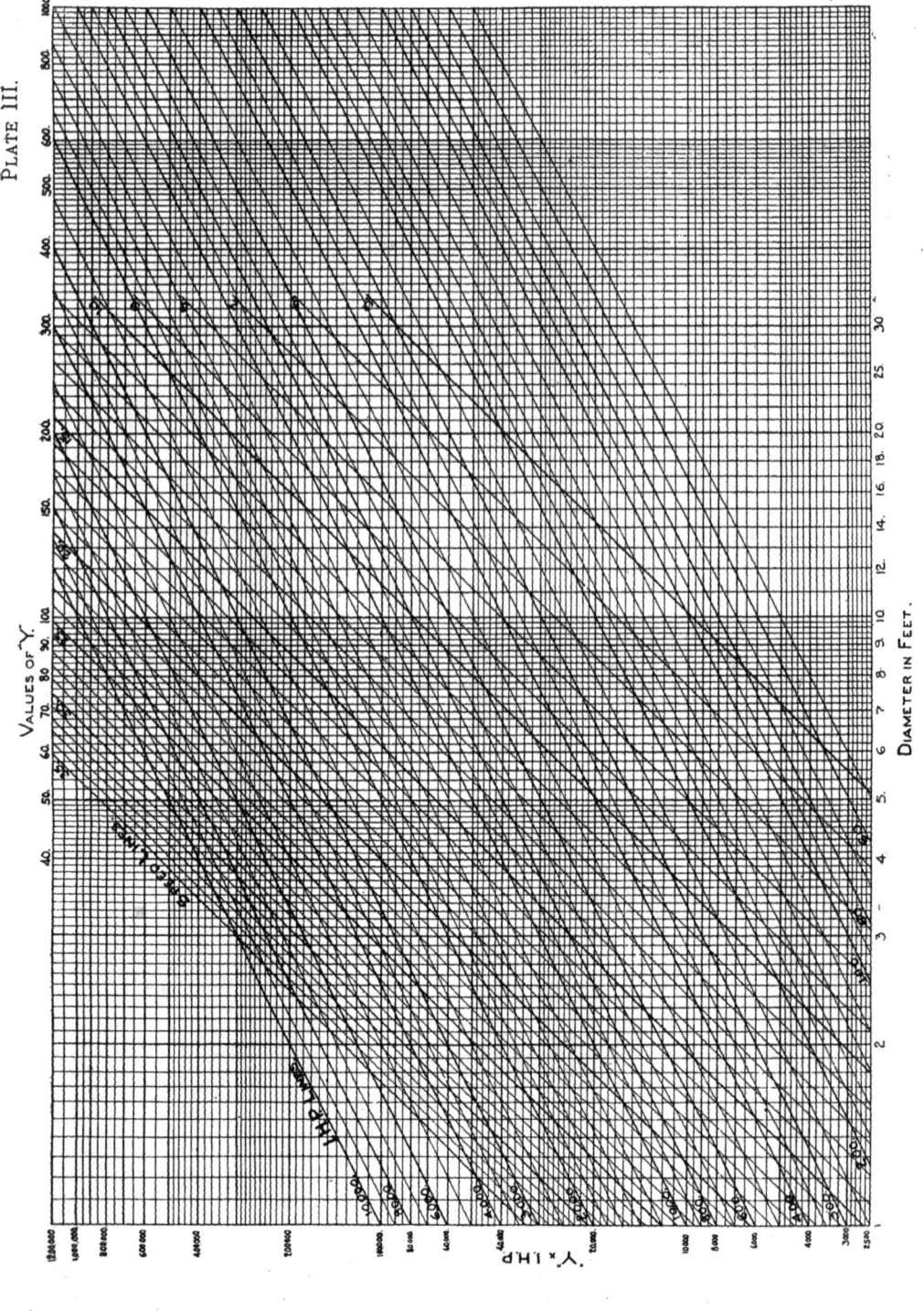

PLATE III.

PLATE I.